METAL CARBONYL SPECTRA

P. S. BRATERMAN

Department of Chemistry,
University of Glasgow,
Scotland.

1975

ACADEMIC PRESS

LONDON   NEW YORK   SAN FRANCISCO

A Subsidiary of Harcourt Brace Jovanovich, Publishers

ACADEMIC PRESS INC. (LONDON) LTD.
24/28 Oval Road,
London NW1

United States Edition published by
ACADEMIC PRESS INC.
111 Fifth Avenue
New York, New York 10003

Library of Congress Catalog Card Number: 73–19001
ISBN: 0–12–125850–5

Printed in Great Britain by
ROYSTAN PRINTERS LIMITED
Spencer Court, 7 Chalcrt Road
London NW1

Preface

This book is addressed to organometallic chemists who, like the author, do not think of themselves primarily as spectroscopists, but who nonetheless realise the need to make the best possible use of spectroscopic data. The mathematical development is descriptive rather than rigorous, and the use of matrix notation is restricted to Chapter 2 and short sections of Chapter 3. Some acquaintance with Group Theory is however assumed.

The greater part of the book is devoted to vibrational spectroscopy, reflecting the central position that this occupies both in routine characterisation and as a source of information about structural features and bonding. Although it would be neither possible nor desirable to refer to all reports of vibrational spectra for metal carbonyls, an attempt has been made to present a range of data extensive enough to act as a basis for comparison with new results.

The final section discusses the use of ultraviolet and photoelectron spectroscopy as well as carbon-13 n.m.r. methods for metal carbonyls. Although the information available is as yet scant, it is not too early to draw attention to the potential usefulness of these methods in metal carbonyl chemistry.

I wish to thank numerous teachers and colleagues, and in particular the research students who have worked with me in this area; Mrs. Frieda Lawrie and her associates for the collection of spectra and for generous help with the literature survey; and finally my wife, Michele, who has not only read and typed the manuscript, and assisted in preparing it for the press, but has made numerous helpful suggestions regarding presentation and style.

Note on Units

With the exception of the reciprocal centimetre as a measure of wavenumber, SI units are used throughout this work. Force constants are measured in Newtons metre^{-1}, and internuclear distances in pm. Some relationships between SI and other units are

$$1\,cm^{-1} = 10^{2c}\,Hz$$

(where c is the speed of light in m. sec^{-1})

$$= 2{\cdot}9979 \times 10^{10}\,Hz$$

$$1\,cm^{-1} = 1{\cdot}9864 \times 10^{-23}\,J, \text{corresponding to}$$
$$11{\cdot}962\,J\,.\,mol^{-1}$$

$$= 1{\cdot}2398 \times 10^{-4}\,eV, \text{corresponding to}$$
$$7{\cdot}4663 \times 10^{19}\,eV.\,mol^{-1}$$

Thus

$$2{,}000\,cm^{-1} = 5{\cdot}9958 \times 10^{13}\,Hz = 3{\cdot}9728$$
$$\times\,10^{-20}J = 2{\cdot}4796 \times 10^{-1}\,eV,$$
$$\text{corresponding to } 23{\cdot}924\,kJ\,.\,mol^{-1}$$
$$\text{or } 1{\cdot}4933 \times 10^{23}\,eV.\,mol^{-1}$$

$$1\,md/\text{Å} = 100\,Nm^{-1}$$

$$1\,\text{Å} \quad = 100\,pm.$$

For a pure harmonic CO stretch, $2{,}000\,cm^{-1}$ corresponds to $16{\cdot}153\,md/\text{Å}$, or $1615{\cdot}3\,Nm^{-1}$. The carbon oxygen distances for free CO (r_e) and for CO in $Cr(CO)_6$ (mean, corrected for riding motions) are $112{\cdot}82\,pm$ ($1{\cdot}1282\,\text{Å}$) and $117{\cdot}1\,pm$ ($1{\cdot}171\,\text{Å}$). [316, 344]

Note on Figures

The spectra shown in this book were generally taken on a Perkin–Elmer 225 Infrared Spectrophotometer at Glasgow University, under normal operating conditions. The help of Mrs. Frieda Lawrie and of Mr. John Black is gratefully acknowledged.

The cell design and related diagrams loosely follow those of commercially available equipment, principally the Beckman-RIIC cells N-05 (Fig. 4.12), F-05 (Fig. 4.15), and VLT-2 (Fig. 4.16).

CONTENTS

Here we are piled in a car leaving the refugee camp
for the safe house in Guinea.

Mariel, Mia, and me (from left to right), with another child,
waiting patiently for our visas after arriving in Ghana.

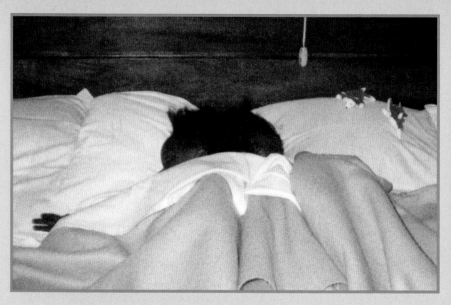

At the hotel room in Ghana, Mia and I slept in a real bed with pillows and blankets for the first time. We snuggled together the way we did on our grass mat at the orphanage, in spite of all the space.

At age four, just a few weeks after arriving in the United States, Mia and I already looked healthier.

Our new brother Teddy delighting Mia and me by tossing us around.

My mother finally let me hold Baby Emma on her last day with us.

Teddy teaching me how to eat spaghetti for the first time.

Papa and me on my first carousel ride at the Jersey Shore.

Teddy took us trick-or-treating in angel costumes
Mama made for us on our first Halloween.

Mia and me, in the turkey dresses that Mama made for us,
with Papa on our first Thanksgiving.

Mia shared her fifth birthday with me because
I had never had a birthday cake before, and she
even let me blow out some candles.

Mama and me riding
the tram to the top of
Mt. Killington in Vermont.

Mama reading a bedtime story to Mia and me, as she did every night.

Mia was good-natured enough to always dance the role of the boy when we performed our version of the party scene from *The Nutcracker*.

At age five, I liked to wear hand-me-down dresses and pretend that I was dancing in a ballet.

Such a happy day! Mia and me kissing Papa when our adoption was finalized.

Mia and me, at age seven, with Mariel and Mama
on our way to see the New York City Ballet.

Mia, Mariel, and I choreographed a dance to perform together for our parents.

Mia, me, and Mariel (from left to right) wearing the blue ribbons
we won in the swimming medley relay.

By age eight, I already wanted to dance wherever I went.
Here I'm at Shenandoah National Park in Virginia.

Practicing arabesques in my bedroom,
wearing my new pointe shoes.

This is my first formal ballet photo.
Luckily, you can't see the stains from the
chocolate bar that melted all over my
tights on the way to the studio.

Papa, Mama, Mariel, Mia, me, and Amie (from left to right) celebrating the finalization of Amie's adoption.

Dancing at the pool after breaking the Tri-County Swim League backstroke record.

Me, at age twelve, with Jestina, Mia, and Bernice (from left to right), singing and dancing while on vacation in California.

Leaping over Mia in the dance I choreographed
to the music from *Pirates of the Caribbean*.

Bernice, Mia, Mariel, me, and Jestina (from left to right)
in our holiday photo, when I was thirteen.

During this photo shoot in Johannesburg, South Africa, when I danced for the Joburg Ballet, I worried that the pigeons would have little respect for the beautiful costume. Fortunately, they left it alone!

Reviewing footage for *First Position*, the ballet documentary I was featured in, with the movie's director, Bess Kargman.

The movie poster for *First Position*.

Mama and me in Los Angeles for my appearance on *Dancing with the Stars*.

Being interviewed on *Dancing with the Stars* by host Tom Bergeron.

Dancemagazine

May 1979 · $2.50

Dancelogue: A Celebration of the Diversity of Dance

Dancing for Pleasure: Kirk Peterson

Triumphant Season: New York City Ballet's Wintering Over

International Calendar of Summer '79 Dance Events

Pennsylvania Ballet's Magali Messac

(top right) The magazine cover that I found as an orphan in Sierra Leone, which inspired me to become a ballerina.

(bottom right) Dancing with the Joburg Ballet.

(left) Dancing the role of the Black Swan in *Swan Lake* with the Dance Theatre of Harlem.

companies other than contemporary ones that had black ballerinas. I found a list of ballet companies in a dance magazine and began my own research. I was determined to find those black women.

A friend of my mother's had once told me that to get into the best black sororities in college, you had to have skin lighter than the brown bags used in supermarkets, and she had failed the "brown paper bag test." I thought about that a lot as day after day I searched through dozens of head shots on ballet company websites, hoping to find a smiling black face. I did find quite a few black male dancers, but rarely did I find a black female dancer, and those whom I did find were light enough to pass the brown paper bag test.

I began to question whether my skin color would prevent me from becoming a ballet dancer. My self-doubts grew when I heard what people said about black ballerinas. That year, in *The Nutcracker*, I danced the part of a Polichinelle, one of the little doll-like figures that pop out from under Mother Ginger's skirt. During rehearsal one of the mothers who were chaperoning us said, "Black girls just shouldn't be dancing ballet. They're too athletic. They should leave the classical ballet to white girls. They should stick to modern or jazz. That's where they belong."

My younger sister told me that she once heard a dance teacher claim, "Black girls can't point their toes."

Once, someone in the ballet world, a man whose opinion meant a lot to me, said to my mother, "We don't like to waste a lot of time, money, and effort on the black girls. When they reach puberty, they develop big thighs and behinds and can't dance ballet anymore." I overheard the remark, but I wasn't supposed to be outside the door listening in, so I couldn't speak up and challenge what he said. My mother did, though, and that made me feel a little better. However, those words still terrified me to the point that I worried endlessly about the fateful day when I'd reach puberty and grow a big butt and big thighs.

I was in an audience watching a performance of *Raymonda Variations* when I overheard a woman criticize a black ballerina, saying, "She'll never make principal dancer; she isn't delicate enough. Black women are just too athletic for classical ballet. They're too muscular. That's why so few of them make it into companies." I flexed my biceps and slid my right hand up my left arm. I gave it a quick squeeze, wondering if my muscles were too big.

Ms. Stephanie, the codirector of the Rock School, made me feel better the next week when she said, "If you keep working hard, I don't see any reason why you can't one day become a world-class dancer."

I thought about that when the mother of one of my dance school classmates said, "Michaela has a lot of strength. She dances like a real brute. Black dancers just have that kind of body."

I cried all the way home that night, at first refusing to

tell my mother what was bothering me. Finally I blurted out, "Do I dance like a brute?"

My mother told me the mean comments that I overheard about black ballerinas were based on jealousy as well as bigotry.

"You need to ignore them," she said.

"But I can't!" I sobbed as I struggled to catch my breath. "I'm worried that I'll never be a ballerina."

My mother's words couldn't comfort me, because I was her daughter and I knew that she would think I was perfect and beautiful, even if I were ugly and too athletic and danced like a brute. Therefore, she couldn't make me believe that the comments were about someone else's prejudice and jealousy and were definitely not true. It took a professional ballerina to convince me of that.

One day I stood outside of my ballet class crying quietly because I didn't want anyone to notice me. We had auditioned for the summer ballet intensive program the week before, and I had skipped a level. A group of mothers went to the director to complain about my placement. Before class they had been whispering together and pointing at me in the lobby. Now I had to face their daughters, and I was afraid to go into class.

My crying wasn't as secretive as I had hoped. As I stood alone in the hallway, Heidi Cruz happened to walk by just then. She saw me crying and stopped. I had no idea if she knew who I was, but she made me explain why I was so upset. "Michaela, you are a very talented young dancer," she said. "And you're going to meet many jealous people. Don't let them take you down. Just hold your head up. Look straight ahead and ignore them, or they will destroy you. Believe me; I know. The same thing happened to me when I was your age."

It had never occurred to me before that I wasn't the first ballerina to suffer this way, and I will always be grateful to Heidi for helping me that day. Knowing that she had experienced the same thing buoyed my spirits. I followed her advice, and I'm glad that I did, because as I grew older, life got harder and more complicated. I didn't need something as trivial as jealousy to bring me down.

Jealousy is a big factor in the world of ballet. There are millions of girls taking ballet lessons and far too few jobs available in companies. The competition to get into these companies is real, and it starts when they are very young. I had to learn to cope with it, or it would destroy my spirit and hold me back from reaching my goals.

I didn't do this immediately. It took a long time for me to learn how to ignore the jealousy and bigotry that had entered my life. At that age I was very confused by the fact that mothers got into their daughters' business. It was hard enough to deal with the jealousy of the other kids without their parents being involved in it too.

It was the parents who often started the meanest rumors. One rumor that spread like wildfire was that my mother had cut three years off my age so that everyone would think that I was very talented. This one upset me more than any other, and I responded to it in my usual way, by bottling it up until I exploded in tears on the way home.

"Why would parents be jealous of me? That just doesn't make any sense," I cried, trying to understand it.

My mother explained to me how some parents try to live through their children's experiences. "If a mother wanted to be a ballerina and couldn't do it, sometimes she tries to live her life through her daughter. That causes her to feel jealousy toward you if she perceives that you are a better dancer than her daughter."

"Will they do it forever?" I asked as I tried to wipe my tears with a crumbling tissue.

She told me that nothing lasts forever. "There will come a day when these girls will no longer be a part of your life, and you'll no longer care about what they once said about you."

Chapter 20

DANCING *THE NUTCRACKER!*

My parents explained to us that when the debils had attacked Mariel's village, they had killed her father and shot off her pregnant mother's leg. As a result of this stress, Mariel was born too early. Her early birth, plus the lead poisoning, malaria, and malnutrition that we all had suffered in the orphanage, had affected her and made it hard for her to learn now.

I was just wild about ballet classes from the moment that I stepped into the Rock School. Mia saw how much I liked it and eventually asked to change to the Rock School as well. Mariel wasn't a bit interested in ballet, but soon after my parents adopted her, she insisted on taking lessons, because she didn't want to wait in the lobby while

Mia and I were in class. So, when I returned to the Rock School at age seven, Mia and Mariel came along too.

When I entered Level 2, my first pointe class, Mia entered Level 1X, the level right behind me, and Mariel entered Level 1. I soon noticed that Mia and I were making real progress but Mariel would fall asleep at the barre and couldn't seem to learn the basic steps and combinations that were taught in her level.

Mariel wasn't slow just in ballet. Despite the fact that she was only four months younger than me and eight months younger than Mia, she was behind us in nearly everything. When we were younger and in the orphanage, Mia and I hadn't noticed this so much, but now that we were older, it was obvious to us.

Even though ballet was so difficult for Mariel, and not much fun for her either, she stuck with it because she wanted to dance in *The Nutcracker* with the Pennsylvania Ballet. You had to be eight years old to dance in *The Nutcracker*, so in my first year back at the Rock School, both Mariel and I, who were seven, couldn't audition.

As I waited in the lobby of the dance school, I looked on with envy as Mia and all of my classmates went up into the dance studio to audition. I learned a lesson just from observing that year, when some kids came down smiling and others came down crying. It's a lesson that I have never forgotten—you don't always get the role that you hope for.

At that time there weren't a lot of little boys taking ballet, so some of the girls had been cast as Party Boys. They were often the ones crying. It seemed that nobody wanted to be a Party Boy.

My stomach twisted into knots as I waited for Mia to come down. She was very tall for eight and her hair was short, and even though I was jealous that she was older than me, I still wanted the best for her. So I worried the whole time that she wouldn't be cast in the envied role of a Party Girl.

When Mia entered the lobby, she was grinning from ear to ear. I raced up to her, and, breathless with excitement, I asked, "What did you get?"

Mia's violet-brown eyes sparkled with joy when she exclaimed, "I'm a Party Boy and a Mouse!"

And so, from my sister, I learned the second half of that lesson—if you don't want the role you got, there's always another dancer who does.

Mia loved dancing in *The Nutcracker* that year. Mama and Papa took Mariel and me to see the show several times, and each time I felt just as excited and longed for my turn. But there were many months of hard work ahead before the next *Nutcracker* audition, and soon the holiday season was over and we were back to work in the dance studio.

The Rock School is what is known as a syllabus school.

That means it has a list of dance steps and combinations of dance steps that you have to learn at each level to move on to the next level. For example, in Level 1, you might be expected to do a combination of dance steps like: tendu to second, relevé, demi-plié, return to first. But in Level 3X you would be expected to do a combination like: fondu front en relevé, close; fondu back, inside leg en relevé, close; fondu outside leg to second en relevé, then plié with the standing leg while the working leg is at forty-five degrees, then go to passé. Repeat in reverse.

Students didn't move up from level to level without being ready. Everybody had different muscle strength, co-ordination, and basic ability, so some kids spent one year in a level and others spent two or three years in a level before moving up.

At the end of that year, Mia and I danced in the Rock School's summer intensive, which was an all-day ballet program, and we'd hurry home at the end of the program to our swim team practice. Swimming had become nearly, but not quite, as important to me as ballet.

In the fall Mia and I learned that we had each been moved up to the next level in ballet, but Mariel had not changed levels. She remained in Level 1, and she was very disappointed because her best friends in Level 1 had moved up to Level 1X without her.

On the way home from class, Mia and I couldn't contain our excitement over our new ballet levels until we heard Mariel sniffling.

"Mariel, do you want to be a ballerina when you grow up?" I asked her. "If you do, I'll tutor you at home."

"Of course I don't!" she answered, with a pout on her face. "I want to be a babysitter."

"Then why are you taking ballet lessons?" I asked.

"Because I want to be in *The Nutcracker*," she answered.

"Well, you're eight years old now, and you can be in *The Nutcracker*. Level 1 kids are the Angels."

"Whew! That's a relief," Mariel admitted. "I really didn't want to move up to Level 1X, because ballet is too hard for me. I'd rather play drums."

On the day of the *Nutcracker* auditions, I knew that Mariel's days as a ballerina were limited. I just prayed that she'd get chosen to be an Angel. I wanted her to have the thrilling experience of performing in *The Nutcracker* at least one time. After her audition Mariel came into the lobby glowing like the star on the top of a Christmas tree. She had been cast as an Angel.

Mia had shot up like a sunflower during the past year, and she knew that she was now too tall to be a child in the party scene. She worried that she'd also be too tall for the role of a Polichinelle, or "Polly," who pops out from under Mother Ginger's dress, and yet she was too young to be cast as a Hoop. "Do you think I'm too tall to be a Polly?" she'd ask me every night before we fell asleep.

I'd answer, "No, I think you'll be a Polly." Then I'd fall

asleep, hoping that I wouldn't be wrong and Mia wouldn't be disappointed.

Thankfully, Mia was cast as a boy Polly and a Mouse. I was cast as a girl Polly and a Party Girl in Cast A. I would be dancing about forty times, half of them with Mia.

I remember how nervous and excited I felt about dancing my first role in a professional ballet. I couldn't even sleep the night before we opened, and I had butterflies in my stomach before I stepped onto the stage. But when the orchestra began playing, I was suddenly calm. I was no longer Michaela. I was a different girl, one of several of Marie's friends who were attending her family's holiday party in nineteenth-century Europe.

That holiday season our lives revolved around *The Nutcracker*. Mia and I listened to the music every day in rehearsal and later in the performance. As though that wasn't enough Tchaikovsky for one season, Mia taught herself to play *The Nutcracker Suite* on the piano at home, and played it endlessly. Mama baked nutcracker sugar cookies that filled the house with delicious smells. From that year on, the sound of the introduction to *The Nutcracker Suite* would overwhelm me with excitement and warm holiday feelings.

Mariel danced her one and only *Nutcracker* performance that year. She did a perfect job of it, despite the fact that one day she ran a fever shortly before the performance and vomited all over the stage while dancing. She was so professional about it that no one even noticed.

However, when those in the roles of Chocolate from Spain danced on after the Angels, one of the dancers slipped on the stage and wrinkled his nose with disgust at the smell.

At the end of the season, Mariel said, "I think I'm going to quit ballet. Being in *The Nutcracker* was a lot harder than I thought it would be."

Chapter 21

TURNING A BLIND EYE

In January 2004 only Mia and I returned to the Rock School. By the following *Nutcracker* season I had skipped a level. I was nine years old and in Level 3X. Mia had just turned ten years old in September, and she was in Level 3. Though we were only four months apart, I looked like a little kid and Mia now looked like a teenager.

During the war in Sierra Leone, Mia had been hit in the head and knocked unconscious. She had no memory of the injury, but it had been so severe that it caused damage to her pituitary gland and made her body grow and mature too early.

She knew that she was now too tall to dance the role of a Polly.

The *Nutcracker* audition was particularly stressful that

year. There had been talk of my being cast as Marie, and I was especially anxious. At the audition I learned that I would not be Marie. *Don't cry in front of everyone, Michaela*, I told myself as I struggled to hold back my tears. I asked permission to use the restroom. There I let the tears flow and quickly wiped them away so that I could return to the audition.

As I opened the door to leave the restroom, two adults passed by. I quickly shut it again when I heard one of them say, "Why not Michaela DePrince? She's perfect for Marie." I nudged it open an inch, just in time to hear the other person say, "Because this city isn't ready for a black Marie." I stood frozen by the door, wondering if the city or the world would ever be ready for a black Marie, or a black Sugar Plum Fairy.

Despite my disappointment, I was excited when both Mia and I were cast as Hoops. But after casting most of the Pollys, the casting director noticed that none of them had experience with that role. Without at least one experienced Polly to lead the way, this dance could become a disaster.

The director looked at his Hoops. I was the youngest and smallest, so he pulled me out of the Hoops and took me aside. "Michaela, I hate to do this to you, but I desperately need you to guide the Pollys. Would you mind being a Polly for one more year? I'll make it up to you next year."

"Not at all. I'll dance whatever role you need me

to dance," I answered with a brave face, even though I wanted to burst into tears. So that year I was a Polly and a Mouse. I reminded myself that I'd be dancing ballet for many years, and there was plenty of time to get a choice role. Now, when I look back on it, I'm glad that Mia got the role of a Hoop, because that was the last time she ever danced in a professional ballet production.

Mia moved into Level 3X in January 2005 and hated the fact that she would need to take classes six days a week. Her true love was her piano, and she worried that she wasn't getting enough practice. I suspected that soon Mia would quit ballet. For years we had done everything together. I dreaded that we might be going our separate ways.

Finally the day came when Mia said, "Mama, I like ballet, but not six times a week. I miss my piano. I want to take piano lessons two or three times a week instead of once."

My heart leaped into my throat when Mia said that. I waited for my mother's response and hoped that she would say, "No way!" Instead Mama asked, "Are you sure?"

"No! She's not sure!" I cried out.

Mia said, "Michaela! What's wrong with you? Mama's asking me, not you. I'm not your identical twin. For me ballet is fun and exercise, until it gets in the way of my piano playing. Then it's not fun anymore." Mia turned to

Mama and said, "I've been thinking about this for a long time, and I'm absolutely sure."

My relationship with Mia was different from my relationship with Mariel or Amie, a Liberian teenager whom my family had adopted when I was eight. Mia and I were much closer. I had assumed that we would do everything together our whole lives. At night, when we lay in bed, we'd plan how we'd marry brothers and buy houses next door to each other. In that moment I saw all those dreams crashing down around me. I felt as if Mia was abandoning me.

"At least we'll still swim together," Mia assured me. And the summer that she quit ballet, it seemed that whenever I wasn't dancing, she and I were at the pool together swimming. When we were eight years old, we had both been part of our swim club's relay team. Together with two of our teammates, we broke a league record in the eight-and-under relay. Now, at ten years old, Mia, Mariel, and I would often take the first-, second-, and third-place ribbons for our age division at the summer meets. That was the summer when I broke two individual league records, one in the butterfly and the other in the backstroke. My coach used to tease that I was fast because I swam with my toes pointed like a ballerina's.

I used to swim without goggles, but the year before I broke these records I started wearing them because I was having trouble seeing the end of my lane. I also began bumping into other students in my ballet class. This happened most frequently when I was doing piqué turns across the room. "Michaela, are you spotting?" my teacher would ask. *Spotting* means continuously keeping an eye on a distant point as you turn. If you don't spot, you turn like a wobbly top. I had learned to spot when I was five or six years old. I was insulted that the teacher thought that, at nine, I couldn't spot.

"Yes, I'm spotting," I answered.

"You don't look like you are," he said. "You're staggering like a drunken sailor as you travel across the floor, and you're knocking your classmates off balance when you bump into them."

That night I tried to spot while crossing my bedroom. I was fine when I turned to the right, but I couldn't seem to spot when I turned to the left. "I think I need glasses," I said to my mother. "My vision is a little fuzzy in my left eye." She made an appointment for me with the eye doctor.

While I waited for my turn at the doctor's office, I admired a pair of Harry Potter frames. "Can I get my eyeglasses prescription filled today?" I asked the optician.

"It depends on your prescription, but we can probably fill it today."

I handed her the Harry Potter frames, and she put them aside for me. Then I was called in to have my eyes examined. Our ophthalmologist spent a lot of time chatting with us, asking about school and ballet, but I wanted her to hurry that day. I was excited about going on vacation to the beach later that afternoon.

The doctor took a look at my eye and gasped. I nearly jumped out of the chair when she did that. Then she called my mother over and showed her my eye through a magnifying glass. Mama gasped too. Then she started to cry. She was always totally matter-of-fact about medical problems, so when she cried I got nervous. "What's wrong?" I exclaimed, thinking that any second I would be crying too.

The vision in my left eye wasn't just fuzzy. It was almost gone. I had some kind of blister on my eye. The doctor thought that I had transferred a herpes cold sore from my lip to my eyeball, but I hadn't had a cold sore on my lip for years. The next thing I knew, I was on my way to the emergency room, without the cool Harry Potter glasses.

The doctor ordered blood tests, and when the results came back we all got a big surprise. I had never recovered from the mononucleosis that I had contracted in Africa when I was four years old. My blood work showed that I still had a very bad infection after five years! The ophthalmologist had never heard of this. She immediately put me on an antiviral drug. The drug was designed to treat the herpes virus. She didn't know for sure if it would work on the Epstein-Barr virus, which was the kind that had caused my mononucleosis, but eventually the blister on my eye healed, and I stopped bumping into my class-mates in ballet class.

My family did get to the beach for vacation that summer. My big brother Teddy drove to the seashore and joined us for a few days. I literally screamed with joy when he showed up. Teddy rode with me on the go-carts and let me steer. Once I had my fill of the go-carts, my sisters and I ran along the beach, watching Teddy parasail by in the sky over the water.

That fall, after a glorious summer, I returned to my ballet lessons at the Rock School. My life was busy with homeschooling, ballet lessons, and *Nutcracker* rehearsals. I didn't worry about my eye too much. Actually, the only times I thought about it at all were when I needed to spot and when my mother dropped medicine into my eye at night. Besides, something much more dramatic was happening in my life that year . . . in all of our lives. It was a drama that was far more serious than my eye.

Chapter 22

LOSS

Teddy was my hero, and often my fellow mischief maker in our family. When Mia and I were very little, he'd play children's songs on the piano for us. When he'd bring his girlfriends home to meet us, he would carry us around on his shoulders and get us all wound up. Soon we'd be dancing and jumping all over the place like wild girls. He'd play circus with us, tossing us into the air and swinging us around.

"Mama! Papa! Look!" we'd shriek with mad delight.

"What's this *Papa* and *Mama* about?" he'd tease. "Do you live in the woods with Goldilocks and the three bears? You need to say *Mom* and *Dad* like me." So, because of Teddy's teasing, we eventually began sounding more like American kids.

Teddy loved to take us to the movie theater. Our parents would always warn him not to buy us soda, but he'd buy us each a huge bucket of popcorn and an extra-large soft drink anyway. He spoiled us, and we knew it.

Teddy taught me how to eat spaghetti in America. He said that I should hold it high above my head and suck one strand down at a time. He even showed me the kind of sucking noise I should make, and he demonstrated how I should make very loud, slurpy sounds when I ate soup in a restaurant.

"Teddy! You're behaving worse than the little kids!" Mom would complain, but Dad just laughed at his antics.

Teddy came to all of our ballet showcases and piano recitals. He attended many of our swim meets, especially the championship meets. When Mia, Mariel, and I were in *The Nutcracker*, Teddy was there. He was the cool, fun brother I never knew I wanted until I was lucky enough to have him. But my high-spirited brother slowly began losing his energy.

By the time he was twenty-three and I was nine, I began to recognize how sick Teddy was. The possibility of losing him took me by surprise, though it probably shouldn't have. Our parents had told us that they had lost two sons to hemophilia, a bleeding disorder. The blood that the doctors had used to treat the boys' hemophilia when they were little had been contaminated with HIV, and they developed AIDS. Cubby died at the age of eleven, nearly two years before I was born, and Michael died nine months after Cubby, when he was fifteen. Even though I saw their pictures on the wall above our fireplace, they didn't seem real to me . . . but Teddy, well, he was real—a dear and precious part of my daily life.

I was used to the idea of Teddy's hemophilia. I often saw him injecting himself with his medicine. To him this was a common occurrence, no different than taking vitamins was to me. Though I had heard the words *hemophilia* and *HIV* many times, I never connected his illness to the thought of his dying.

In the fall following our great summer at the beach, Teddy grew weaker and weaker. Soon he was barely able to walk. He was twenty-three years old and owned his own house by then, but he'd often sleep overnight at our home in the spare bedroom on the first floor. Often he'd spend the day resting on our huge leather sofa, just so that he could watch us play and listen to our conversations, interrupting often to tease us.

Our noise never bothered him. Teddy would claim that we were so funny that we made him feel better. Usually he fell asleep amid our chaos. He was terribly sick for a year.

On November 13, 2004, the day my new sister Amie's adoption was finalized in court, we went out to dinner, but Teddy was too weak and tired to join us. On our way home Mom's cell phone rang. It was Teddy, and I thought that he was calling to congratulate Amie. I expected Mom to pass the telephone to her, but she didn't. Instead I heard her say, "Okay, we're almost home. We'll drop off the girls, and then Dad and I will drive right over to get you."

"Are you picking up Teddy and bringing him over? Is he going to celebrate with us? Will he sleep over tonight?" I asked, but Mom didn't answer me. Instead she ignored my questions and continued talking to Teddy on the phone in a soft, quiet, calming voice the entire way home.

When we pulled into the driveway, Mom said, "Teddy's sick. Your dad and I are going to take him to the hospital. You girls go inside and change into your pajamas. You can watch television until we get home."

My dinner did somersaults in my stomach. "Can I come, please?" I begged.

"Can I come too?" Mia asked.

"No one is going except your mother and me," Dad said.

Amie, Mia, Mariel, and I sat cuddled on the sofa, trembling and whimpering like scared kittens. We were too worried and distracted to watch TV as we waited for hours to hear from our parents.

Finally the telephone rang. Mia and I ran to answer it. We pressed our faces close together so that we could both hear. "We're on our way home, girls. We'll be there in ten minutes," Mom said.

"How's Teddy?" we both asked at the same time.

"Is he coming home with you?" Mia asked.

"Do you want us to turn on his electric blanket to warm up his bed?" I asked.

"No," Mom said.

"Is he okay?" I asked.

"We'll talk about it when we get home," Mom answered. "We're almost there."

Mom and Dad put their arms around us when they got home, and led us to the living room. Then, in a soft, shivery voice, Mom simply said, "Girls, Teddy died tonight."

I thought that my heart would break into a million pieces. I heard a loud scream in my head and realized that it was my own voice. All of my sisters screamed too. Our cries were loud, howling wails. Mariel rolled on the floor, hysterical with anguish. Mia and I clutched each other desperately. Amie threw herself into our mother's arms. Mom and Dad reached out and filled their arms with all of us, holding us close in a circle of comfort and love. It seemed to me that my parents had octopus arms, because they were able to wrap all of us together, but they couldn't hold our broken hearts together.

That day I felt such a pain in my chest that I thought my poor broken heart was bleeding. I had never before felt that sad, not even in Africa. But perhaps I was too young to remember exactly how I felt in Africa. Or perhaps I hadn't understood back then, when I was a tiny child, that death was forever.

Now that I knew that death could snatch away my loved ones even in America, I began to worry about everyone in my family, especially my parents. The thought of losing them terrified me. I was so afraid that I would lose them that I tried to pull my affections away from them, reasoning that then it wouldn't hurt me as much when they died.

What was worse, I pulled away from my sister Mia as well. Mia had been the one constant in my life for many years. She had been my first real friend and only real ally in the orphanage. Yet now I rejected her too. We began arguing constantly, and I stopped confiding in her as I used to do.

I learned from losing Teddy that people have different ways of dealing with grief. I know now that I chose a painful way of coping with his loss. Amie and I rebelled and held all of our loved ones at arm's length, while Mia and Mariel drew closer within the circle of family.

In the time after Teddy's death, I didn't really understand my feelings. My mom tried to help me and even insisted that I see a therapist to sort out my feelings, but I refused. Unless I was carried kicking and screaming out of the house, there was no way anybody would get me to see a therapist.

I was driving my parents crazy, but they never gave up on me. They never stopped loving me, even when I rebuffed their affection. Even in my mixed-up state of emotional confusion, that brought me comfort.

I think that there might have been a second reason why I was angry with my parents. I felt that they hadn't kept Teddy safe, which of course wasn't at all true. They had done as much for my brother as was humanly possible.

But I felt that if they couldn't keep him safe and well, then how would they protect me? For a very long time after he died, my anger over this got me into situations from which I needed to be rescued . . . perhaps to prove to myself that my parents would save me, even though they hadn't been able to save Teddy.

•• *Chapter 23* *•*•

MOVING ON

E ven though our home was filled with so many noisy
girls, it seemed much sadder and lonelier without
Teddy. Mia, Mariel, Amie, and I had begun knitting when
I was eight years old. I was nearly ten when Teddy died,
and we were still knitting. We'd sit together at night, knitting and chatting before we went to bed. A couple of days
after Teddy died, I held up a soft, thick gray scarf that I
had been working on and asked, "What should I do with
this? I was knitting it to go with Teddy's favorite hoodie. I
wanted to give it to him for Hanukkah."

"Maybe you should finish it and give it to Daddy," Mia
suggested.

"No, it would make him too sad," I said. "And I can't

give it to either one of our other brothers. They know I was knitting it for Teddy."

"Then maybe you can finish it and give it to your favorite dance teacher," Amie said.

"No, it's Teddy's. I can't do that," I said. So instead I just kept knitting and knitting. I took out the scarf whenever I felt sad and knitted more and more of it until it was nearly as long as my bedroom. It was my way of dealing with the sorrow of missing Teddy. And I probably would have knitted that scarf forever if the store hadn't run out of matching yarn.

Less than two years after Teddy died, my family decided to move to Vermont. My parents had lived there when they were first married, and they said, "All of our memories of Vermont are happy ones. We've lost three sons in New Jersey. We have too many sad memories here."

My sisters and I agreed that a change might be good for all of us, especially because our parents decided that instead of homeschooling, Mia, Mariel, and I could start sixth grade in a regular public school.

On our first morning in Vermont, we ate hot, fluffy pancakes with real maple syrup, something we could rarely find in New Jersey diners. Then we searched for a new home, one free of sad memories.

Soon we found the perfect home in Williston. It was close to the new dance school that I would attend and only a short distance from a piano teacher for Mia. When we went to visit it, I spotted a picturesque library standing close to Williston Central School. Our parents agreed that we could visit the library while they saw the nearby house.

At the library I met a group of friendly girls. They introduced themselves to my sisters and me. Some of these girls would be in my sixth-grade class that fall if we bought the house.

The girls told us how lucky we'd be, because the house was on the bike path. They told us that depending on the season, we could walk, ski, snowshoe, or ride our bicycles back and forth between the school and its playground, playing fields, and ice-skating rink. I began to look forward to the fun I would have there, even though I was still sad to leave my friends and my ballet school behind.

Every time I felt sad about moving, I'd think of how painful it was to celebrate birthdays and holidays in our New Jersey home now that Teddy was gone. Every room was full of memories of him. He had such a huge personality; he had especially filled the kitchen, dining room, and family room with his smiles and laughter. The mall, the parks, the movie theater, the restaurants . . . his memory was everywhere I went. We all talked about this, and I knew that it was important for my family to start fresh somewhere else.

•◆•

Before we moved to Vermont, I came up with a plan to leave my mark on New Jersey. I decided that I wanted to break the Tri-County Swimming League's fifty-meter girls' backstroke record for my team.

When they called my race at my last meet, I jumped into the pool, gripped the edge of it with my fingers, and curled my legs in front of me with my feet pressed firmly against the wall. I shot off the wall at the bang of the starter's pistol and slipped headfirst under the water. I came up stroking hard and fast, and I could hear the crowd cheering.

When my fingertips touched the wall, I stood and looked around to see who had won. I burst into tears when I discovered that I was the only one standing there. I had beaten the next swimmer by a couple of body lengths. Those cheers had been for me. I had succeeded in breaking the record! That moment of triumph was my final memory of my life in New Jersey.

Even though I loved swimming and the thrill of competition, it didn't make my heart soar as ballet did. Ballet was taking up more and more of my time, so it was getting harder and harder to find the time to fit in swim practice.

Also, I was petite, compact and small-boned. On looking at X-rays of my knee and wrist, a doctor had once predicted that I would grow to a full height of sixty-three to sixty-five inches. That would be a wonderful height if I aspired to become a prima ballerina, but it wasn't tall enough if I wanted to become an Olympic swimmer. I began to notice that more and more of my swimming competitors towered over me, standing several inches taller and outweighing me by about fifty pounds each. By sixth grade I was still wearing children's sizes in clothes, and they were nearly full-grown women.

I had to face the fact that I did not have the body of an Olympic swimmer, and I would not grow into one. In a way I was relieved because now I could dedicate myself to ballet without feeling too guilty about quitting swimming. And after all, I was absolutely passionate about ballet!

Chapter 24

GROWING APART

"Why don't you stand up for me? Why do you always side with your new friends?" Mia cried as we walked home from school one day. "*Your friends* are mean to me! *You* are mean to me!" she shrieked as she hobbled with blood pouring from the cut on her knee. When she had fallen on a rock, a group of kids had laughed.

"My friends are *not* mean!" I shouted back at her. "You're the one who's mean. You ignore them when they try to be nice to you."

"Nice? Nice?" my sister sputtered. "They're not trying to be nice. They make fun of me and talk about me when I can hear them."

We hadn't been going to our new school for very long before it became evident that Mia and I were going our different ways. We had chosen to be in the same home-room, or "house," as they called it at our school, because we had always been the very best of friends. I was getting to be more popular, though, because I could do things like a grand jeté and because the other kids thought that whatever I said was funny.

Mia, on the other hand, was shyer, quieter, and much more serious than me. She was every bit as talented as I was, but in the eyes of sixth- and seventh-grade girls, classical piano and oboe weren't nearly as cool as ballet. Also, for a reason that I didn't understand, the most popular sixth-grade girls took an instant dislike to her. Now I realize that it was probably because she is gorgeous. The boys drooled over her, even though she wasn't very interested in them then.

Mariel was in a different house at school. She had different friends, and as a special-education student, she was more protected by the staff, who looked out for the kids with learning differences. But no one looked out for Mia in school, not even me.

Mia had defended me and protected me when we were in the orphanage in Africa, but when she needed the same protection, I wasn't there for her. I am sad to admit that at the age of eleven, it was more important to me to be popular than to be a good friend to my sister.

At that time I didn't realize how much she was suffering. The boys were making inappropriate comments to her, and I ignored the situation. She finally got up the courage to complain to the dean of our school. The boys were called down to a meeting with him, and they admitted to their wrongdoing.

This got the boys off Mia's case, and many of them then befriended her, but now the girls *really* didn't like her since she had gotten some of the most popular boys in trouble. Looking back on this, I realize that this was all typical middle school angst, but I was just a typical middle school kid, so I was acting like one.

Once again I sided against my sister, complaining about her getting *my* friends in trouble. I know now that what I should have said was "I've got your back, Mia." Just as she had had my back in the orphanage.

For Mia the first semester of middle school was a nightmare, but for me it was a delight. I was doing well in my classes. I was having a wonderful time and making friends at my new ballet school. When the cast list was posted for the *Nutcracker* production, I was overjoyed to see that I had been cast in seven roles. Then, on a snowy evening in January shortly before my twelfth birthday, I fell down the icy steps of the dance studio, and everything changed.

I had been used to the ways of the Rock School, where the teachers didn't want an injured student dancing. There, an injured dancer would watch class until she felt well enough to dance again. I asked for a note from my mother and returned to my new class the next day. The teacher glanced at the note, grunted, and shoved it into his pocket. Then he pointed to the barre and commanded,

"Dance." With great effort I lifted my leg, but it went only as high as the lower of the two barres. The teacher walked by, grabbed my leg, and pulled it up to 180 degrees—until my toes pointed to the ceiling.

I was in so much pain that I dropped to the floor, squealing like an injured puppy. "Here, we work through injuries," the teacher said. I could barely walk to my mom's car that evening, and I never returned to that dance school again. Suddenly I no longer had a place to dance ballet.

As lost as I was without my ballet school, good things did happen as a result of my injury. I actually began to grow closer to Mia again. We began taking classes together at a new dance studio that had opened nearby. A French-woman who had studied ballet at the Paris Opera Ballet School owned it. Though Mia wasn't in my ballet class, she took jazz and modern with me. We had so much fun. After that we began confiding in each other again, and she forgave me for treating her so poorly at school.

Without telling my mom or me, my ballet teacher applied to the Dance Theatre of Harlem's summer intensive on my behalf. One afternoon, as Mia and I were rushing to head off to dance class, the telephone rang. I heard my mother say to the person on the telephone, "Yes, I'm Michaela's mother."

"Who was it? What did they want? Was it about me? Was it my teacher? Am I in trouble?" I asked when she hung up, trying to remember if I had done anything that would require my teacher to call home.

"That was the Dance Theatre of Harlem. You've been offered a full scholarship to the summer intensive there!" my mother exclaimed.

I was overjoyed, but I also felt a little dismayed. Because Mia and I had grown close again, I wasn't at all thrilled about spending seven weeks of the summer away from her.

"Can Mia apply to the summer intensive there?" I asked.

"Michaela!" my sister shouted. "I don't want to go to ballet summer school. I'm going to jazz camp with my friends from band. Though it might not be as glamorous as an overnight program in New York City, it's important to me. I don't want to miss it."

"But who will talk me to sleep?" I asked, thinking of the thousands of nights that Mia had either sung to me or told me stories until I fell asleep.

"You'll probably have a roommate to talk to," Mia said, and she was more correct about that than she could have imagined.

When I arrived at the youth hostel where the Dance Theatre of Harlem summer intensive students stayed, I discovered that I had seven roommates and one counselor in my room. My problem that summer was not needing someone to talk to me or tell stories at bedtime. It was having everyone quiet down so that I could sleep!

While living in the hostel, I learned that hair extensions were a very important part of the other girls' lives. I sat through many a long night as my new friends tugged and braided to give me long hair. "That's the problem with having a white mother," one of the girls said as she braided in my extensions.

"What is?" I asked.

"She doesn't know how to braid your hair."

I wanted to defend my mother, but at the same time, she had never braided my hair or that of my sisters. *Can it be true that white mothers can't braid hair?* I asked myself. I felt like calling and asking her, but I worried what her answer might be. The color of her skin had never been an issue for me. I didn't want that to change.

Besides hair extensions, the DTH introduced me to something else that year. I had never danced in a class of predominant black dancers before. Though there were many black students in the beginning levels at the Rock School, most of the students in my level had been white, and we all wore pink pointe shoes and pink tights, like ballerinas the world over.

I had learned from a former teacher that wearing pink tights is supposed to give the body the illusion of greater length and extension. But pink tights don't have the same effect on a black dancer as they do on a white dancer. When I wore pink tights, I was cut in half—shortened—

and I was already short enough. I was at a disadvantage when dancing in a room full of white ballerinas. They looked long and slim. I looked short and squat.

At the DTH we were required to wear dance shoes and tights to match our skin. In the case of most of the girls, including me, that meant brown. My particular shade of brown was Fashion Brown, and it came in a spray can. I sprayed all of my pointe shoes that color and dyed my tights the identical shade of brown. It was uplifting to be able to wear brown pointe shoes and brown tights. I had never felt so long, lean, and elegant before.

During the summer that I danced at the Dance Theatre of Harlem, the legendary Arthur Mitchell was still the director there. He was an imposing figure with a booming voice and strong presence, which made him look much taller than he really was. Though I had tremendous respect for him and eventually grew fond of him, Mr. Mitchell's voice frightened me in those days. It was what my sisters and I called an African voice, and it reminded me of the voices of the authority figures in my life, like Uncle Abdullah, Papa Andrew, and the terrifying debil leaders of the RUF. Though I had been in the United States for several years, I still quaked in fear at the memory of them. Whenever I heard Mr. Mitchell's voice, I'd dash around a corner or into the ladies' room to hide from him.

Except for his voice, everything about Mr. Mitchell

was wonderful. He had an incredible ability to teach. The previous spring I had gone to him with my contemporary dance routine for the Youth America Grand Prix competition. He watched it and said, "It's pretty good, but let's make it better." He then showed me some tiny nuances: a lift of my chin, a tilt of my shoulders, a flex of my wrist, and a brisk stomp of energy in my feet as I stepped flirtatiously in my pointe shoes. In just minutes he turned good choreography into a great dance.

During the summer intensive Mr. Mitchell was always in the middle of everything. On a late afternoon in the studio, the heat might be causing everyone to droop when suddenly he would appear, and like Drosselmeyer in *The Nutcracker*, he would add a flash of magic to the room. Spines would straighten, heads would lift, and eyes would sparkle. Everyone would become instantly alert, because nothing escaped his notice. A sickled foot, a bent knee, a raised shoulder, or a turned-in hip would draw his attention, and his booming voice would call out your name, causing your pulse to race. Mr. Mitchell had dedicated much of his life to bringing ballet into the lives of Harlem's children, and he always managed to share his thrill and passion for ballet with his students.

Mr. Mitchell nicknamed me Mickey D and Miss Sierra Leone. Oh, how I hated those names, but I tolerated them because I knew that he used them affectionately.

He believed that I showed promise, so he coached me privately after my regular classes ended. He even gave me a solo at the DTH street festival. Crowds of people lined the street that day. I was so thrilled to be chosen for the solo, but I nearly fainted with embarrassment when Mr. Mitchell introduced me to the crowd on the street as Miss Sierra Leone.

Despite his voice, Arthur Mitchell was the first black man that I grew to trust.

Mia and our parents picked me up from the summer ballet intensive in August. On the way home my mom turned in her seat to tell me something and suddenly cried out, "Your hair! Your head! You have bald patches!"

I panicked and began to unbraid my extensions. Mia helped me. It took us about four hours to get them out. When we got home, I washed and conditioned my hair. Then I looked in the mirror. Tears filled my eyes. I wanted braids.

The next morning I asked my mother, "Why can't white women braid in hair extensions?"

"I don't think that has anything to do with the color of a woman's skin, Michaela," she answered.

"Well, can you braid in hair extensions?"

"Of course I can," my mother answered.

"Then why didn't you ever do it before?" I asked.

"You never asked for them before."

When Mom said that, I felt a flood of relief. There was

nothing wrong with a black girl having a white mother, after all.

I gave my scalp a couple of weeks to rest, and then my mother braided in hair extensions for me. They were small and fine and didn't tug at my scalp. When she was done, I had two or three hundred microbraids in my hair. It took her over twenty hours to do my entire head, but they looked like my real hair.

Soon black women began stopping me on the street. "Where did you get these beautiful braids?" they'd ask.

"My mother did them," I'd proudly tell them.

If she was around, I'd make a point of having them meet her. It tickled me to see the looks of astonishment on their faces when they saw that my mother was a white woman with blond, poker-straight hair.

That summer ended in a burst of glory. Mia and I were intent on mending our relationship while we were on a family camping trip. All the conflicts of our sixth-grade year seemed to disappear during those magical days of kayaking on the lake and nights of toasting marshmallows by the fire. There is something so indescribably special about giggling and sharing confidences in a tent under a starry sky. Mia and I slept side by side, just as we had in the early years of our lives.

·•· *Chapter 25* ·•·

THE YOUTH AMERICA GRAND PRIX

When I was younger, I watched with envy when the teenagers at the Rock School participated in the Youth America Grand Prix, or YAGP. The YAGP is the largest international ballet scholarship competition in the world. I too wanted to dance a solo onstage, wearing a real tutu like the ballerina on the magazine cover that I had found years before.

In the fall of 2005, when I was ten, Ms. Stephanie at the Rock School had finally asked me if I would like to compete in the YAGP. Of course, I jumped at the chance.

I was the first student from the Rock School ever to compete in the Pre-Competitive Division, a name that is truly an oxymoron if there ever was one, because I soon learned that this youngest division was *extremely* competitive. I wanted to do especially well so that other young dancers from the Rock School could one day compete in this division too.

One of my competition dances needed to be a contemporary one, choreographed just for me. The other dance needed to be a variation from a classical ballet. You might get away with making mistakes in the contemporary variation, even forgetting your steps, because the judges had not seen the dance before. However, you had to dance the classical variation perfectly, because all of the judges were professional dancers. They knew every step of every classical variation.

Natalya Zeiger, whom we called Ms. Natasha, was my classical coach. For my classical variation Ms. Natasha chose the Gamzatti variation from the wedding scene in act three of the ballet *La Bayadère*.

I adored Ms. Natasha. She had been trained at the Bolshoi Ballet Academy and danced as a soloist at the Bolshoi Ballet before coming to the United States. In my eyes she was the epitome of a Russian ballet dancer: tall, slender, elegant, and graceful. I was eager to please her.

Ms. Natasha was firm but patient as she coaxed me through the steps. The version of the Gamzatti variation that I was performing was full of développés à la seconde. For each of these I had to move a leg up through the retiré position, in which my leg was bent and my toe was pointing to my opposite knee. Then I had to lift the leg up and unfold it to the side until my foot was high in the air and my toes pointed to the ceiling. These took a lot of concentration, control, and hard work. If I did them wrong, it would just look as if I was kicking my foot to the side and high into the air without any grace.

Ms. Natasha was also a very kind person. She invited me to her home and spent hours teaching me how to apply stage makeup. At ten I couldn't imagine a more exciting experience than having a real ballerina teach me how to put on makeup. To this day I follow Ms. Natasha's advice about eye shadow and lipstick. I can still hear her voice telling me to spread the light eye shadow up to my eyebrow, and to apply the darker shade below that. "And remember, Michaela, when you put on the eyeliner you want to go a little past the outside edge of your eye, so that your eyes look larger."

Lipstick was a problem for me. I have very full lips, and I had to be careful not to spread my lipstick beyond their edges. Ms. Natasha showed me how to use a lipstick liner before applying the lipstick, so that it was like coloring in the lines.

Ms. Natasha had a whole collection of tutus that she lent her students, but they were too large for me. The YAGP doesn't require fancy or expensive costumes, and officially it states that one is not required, but I knew that all of the competitors wore beautiful costumes, and I wanted to do the same.

My mother and I spent hours online, searching for tutus to order or rent. Since this was my first ballet competition, we had no idea that tutus and custom-made costumes needed to be ordered months ahead of time. Even if time had not been a factor, the prices of the tutus were too high.

"Seven hundred and fifty dollars! Nine hundred and ninety-five dollars! One thousand one hundred and fifty dollars! Oh, sweetie, we can't afford these, but maybe I can make one for you."

"Maybe we can rent one?" I suggested.

Mom shook her head and frowned. "It costs a hundred dollars a day to rent, and the deposit is hundreds of dollars. By the time I pay the deposit and the rental fee, especially if you make it to the finals in New York City, the cost will be nearly equal to buying one."

"But, Mom, you'll get the deposit back when you return it," I said, dreading the thought of wearing a homemade tutu.

"That's if you don't spill on it, rip it, or lose it," she said.

I didn't think that was funny at all, but my mother was right. I was graceful in the ballet studio, but elsewhere I was always tripping and running into things. And I was constantly spilling food on my clothes.

I shrugged and sighed. "Okay, so how are you planning to make this tutu?"

The very next day my mother dropped me off for my ballet classes and drove to the fabric district of Philadelphia. That evening, when she picked me up, I found bags of puffy white tulle and trimming on the backseat. In one of the bags I discovered pale gold organza with iridescent gold sequined designs. "Wow! This is beautiful!" I exclaimed.

"That's your tutu fabric. Now I have another surprise for you," Mom said as she passed back a smaller bag. Inside was a flat, square box. "Open it carefully. You don't want to drop what's inside," she warned.

When I opened the box, I let out a cry of delight. Inside lay a handcrafted, gold-plated tiara trimmed with crystals. It was perfect!

I loved the Gamzatti variation simply because it was the first variation that I did and I had nothing else to compare it to. Performing it made me feel majestic and free, until the night of the competition. Then I felt jittery with worry because the Rock School had held a compulsory meeting for the YAGP participants, and I had barely enough time to get to the YAGP venue afterward. To reach the stage in time, I had to change into my costume in the backseat of my parents' minivan. I arrived, trembling, about six minutes before my name was called.

This was a complicated variation, involving technically difficult and complex combinations of steps. The version that I was doing included several en dedans turns in attitude to arabesques in plié, which required intense focus on my balance while I maintained a pleasant expression on my face. I also had to cross the stage in a series of développés à la seconde. How could I possibly do this in my present condition? *Breathe deeply. Relax, you're not a kid. You're Gamzatti, an Indian princess, and it's your wedding day,* I told myself as I stood in the wings.

Finally I heard the announcement: "Michaela DePrince, age ten." The music began, and I stepped onto the stage. I felt beautiful in my glittering tutu and tiara. For the minute and a half that the variation lasted, the difficult steps that I had agonized over as Michaela, ballet student, came easily to the glamorous Gamzatti. I lost myself in the role. It was an awesome feeling!

I won the Hope Award in Philadelphia that year. It was the top honor for dancers in my age division.

··· *Chapter 26* ···

A HOMEMADE TUTU

I wasn't sure if I'd be allowed to compete at the YAGP
when I moved to Vermont. The director of my first
ballet school in Vermont didn't believe in ballet competitions. However, after my injury caused me to change
studios, I discovered that my new teacher was more than
willing to help me prepare for the 2007 YAGP. She had
wonderful choreography skills and used them to help me
get ready for the YAGP in a hurry.

Again my mom sewed my contemporary costume, but
it was impossible to sew a tutu on such short notice, especially since I didn't even know for sure what variation
I would dance. "May I dance the role of the dying White
Swan?" I pleaded with my ballet teacher.

"Oh, Michaela, that role requires not only adult artistry, but a lifetime of heartbreak," she answered with a gentle smile.

"Then may I do the Black Swan?" I asked.

"Michaela! The Black Swan is a seductress and you have just turned twelve."

"Well, what's a seductress?"

Now my teacher laughed outright. "That's why you can't be the Black Swan. You don't even know what a seductress is. Suppose you dance the role of Paquita— I think that variation number six, the jumping girl variation, would be perfect for you. You will come flying out onto the stage in a series of five grands jetés, and I know how you love that."

Now I laughed too. The grand jeté had always been my favorite dance step. I loved leaping high and far. It gave me the sensation of taking flight! That afternoon I watched a video of my Paquita variation with my teacher, and then, after I warmed up at the barre for forty-five minutes, we worked on it for an hour and a half.

My favorite part of this variation began eight seconds into it, when I did the five grands jetés, which made me feel as if I was flying. Then I came down to earth, and the tough part began. Tours jetés followed. These are turns within leaps. I knew that I could lose my footing doing them if I wasn't very precise about landing. These turns were followed by arabesques into attitude turns, finishing with a double en dedans pirouette, which I found to be the hardest part. I left that first rehearsal drenched in sweat, feeling as if it was midsummer. I came to my senses fast enough when I walked out into a below-zero winter snowstorm.

While I was worrying about the dance, Mom was worrying about the costume. "Michaela, what does Paquita wear in the variation you're dancing?" she asked the next day.

"A burgundy tutu with gold trim," I answered.

My mother's eyes grew wide with horror. "Michaela, you'll disappear onstage!"

I laughed. She was right. Burgundy on my chocolate skin, against the navy blue background of the YAGP stage, would make me disappear. I'd look like two eyes and flashing white teeth. "Maybe we can reverse the colors," I suggested.

Mom dashed out of the house and returned a half hour later with several boxes of yellow fabric dye and small bottles of burgundy fabric paint. We poured the dye into the bathtub and turned on the Jacuzzi to mix it.

We dunked the white tutu that I had worn the year before into the yellow dye and swished it around. I had no idea how heavy a wet tutu could be until I tried to lift it out. "Mariel! Help us!" I shouted.

Mariel is probably the strongest girl I've ever met. If she wanted to, she could probably compete in weight-lifting championships, but she didn't need to lift weights that day . . . just my tutu. Mariel got a grip on the bodice of the tutu and squeezed the excess water out of it. Then together we stuffed it into a plastic trash bag, giggling the whole time. Finally we carried it downstairs and dumped it into the dryer.

My once-fluffy-and-delicate tutu sounded like a bunch of tennis balls as it thumped around in the clothes dryer. I sat in the kitchen, fearing that it would turn out a tangled mess. Eventually the loud thumping stopped, replaced by a gentle swishing. When I opened the dryer, out popped a perfect golden tutu, as fluffy now as it had been before the dye job.

With my mother's help, I painted the flower petals in the lace a deep burgundy. I had a perfect Paquita tutu when we were done. I was ready for the competition.

It was bitter cold and snowing in Vermont when we left for the competition in Connecticut. It was a harrowing drive, and my mom tripped in the hotel, hurting her shoulder, but we made it on time, where I was quickly swept up in the events of the day. I was so excited to see old friends from Philadelphia and New York City there, and to make new ones from all over the East Coast.

My first variation was my contemporary, danced to exotic Middle Eastern–style music. My mother had sewn a beautiful scarlet-and-gold brocade and gauze costume, trimmed in multicolored crystals. My teacher had wanted me to enter the stage dancing while wearing a thin red veil draped over my face. "Can you see through the veil?" she'd asked me over and over again during rehearsal in her brightly lit studio.

"It's perfectly okay," I'd reassured her as I rehearsed. By the day of the competition, I knew the variation so well that I could dance it blindfolded, which turned out to be most fortunate for me.

When my music began, I dropped my veil over my face and danced onto the stage. In the darkness of the theater I couldn't see a thing! I probably should have ripped the veil from my face, because I was worried at first that I'd fall into the laps of the audience, but suddenly I got into the mood of the music, and I was no longer Michaela. I was a Bedouin dancing girl. It didn't matter that I couldn't see the stage or the audience, because I was no longer performing at the YAGP. Instead I was dancing in a sheik's tent, swaying along with the silk hangings blown by the wind.

Finally I reached the musical note that signaled the removal of the veil. I ripped it from my head, and the Bedouin tent disappeared. I was Michaela again. My many hours of rehearsal had paid off. I was not teetering on the edge of the stage.

When I raised my head, I saw hundreds of people sitting mesmerized. They must have realized that I couldn't see through my veil, because at the end of the performance everyone cheered wildly. I grinned as I curtsied, and then ran off the stage.

After the two-day competition, I slept most of the way home, so I arrived full of energy. I was busy telling my sisters about everything when I heard my mother say to my dad, "Would you mind driving me to the emergency room? I can't lift my arm to steer. I think that I broke my shoulder when I fell in the hotel."

I lay awake all that night, waiting for her to get home from the hospital. The last time someone in my family had gone to the emergency room, he never came home. My stomach was in a tight knot by the time I finally heard the ice in our driveway crackle as our car pulled in at five o'clock in the morning. I was on the verge of tears when I heard my mother's voice in the kitchen, but I tried to play it cool when I saw her standing there with her arm in a sling.

Mom's shoulder recovered in time for her to come with me to the 2007 YAGP finals in New York City. As soon as we got home from that event, I began thinking about my variations for the 2008 YAGP. Once again I lobbied for one from *Swan Lake,* and once again my teacher cast me as a princess. But this year I would dance the role of the ultimate teenage princess—Princess Aurora from *The Sleeping Beauty.* I would dance her sixteenth-birthday variation from act one.

This year I would be in the YAGP Junior Division, and a tutu with a boned bodice was in order. Mom and

I looked online for a costume. Of course, prices had gone up.

I considered dyeing my old tutu, but Sleeping Beauty wore a rose-pink costume in act one, and if I dipped my gold tutu in pink dye, it would probably turn a weird shade of pumpkin. Besides, when I tried it on, I realized that I had grown a lot. For years Mia had towered over me. Now I was taller than her, and I couldn't even tug my tutu over my chest.

"What am I going to wear for a tutu?" I asked.

"Let me worry about the costume, Michaela. You worry about the dancing," Mom said.

·•· *Chapter 27* ·•·

MY FIRST TOUR

While my mother was planning my YAGP costumes that year, Ms. Madeline, a real-life fairy godmother, appeared and gave me a special gift. Madeline Cantarella Culpo is the director of the Albany Berkshire Ballet, or ABB, a small regional ballet company with studios located in Albany, New York, and Pittsfield, Massachusetts. It brings its ballets to small towns and sparsely populated areas of Massachusetts, upstate New York, Vermont, and Canada, where the residents would rarely, if ever, have the opportunity to see professional ballet without the ABB.

The ABB tours *The Nutcracker* every year, using the services of a handful of professional ballet dancers, a few teenage apprentices, and dozens of children from the areas it serves. The same ballet dancers and apprentices dance throughout the tour, but the children change from town to town.

When the ABB arrived in Burlington, Vermont, I auditioned for one of the child roles.

On that chilly morning in September, I arrived at the New England Ballet Conservatory expecting to be cast as a Soldier or a Rat. But my teacher had other ideas in mind. She spoke to Ms. Madeline about possibly taking me on as an apprentice.

After the audition Ms. Madeline took me aside and invited me to learn a variation from "Waltz of the Flowers," one of the professional dance roles in the *Nutcracker* ballet.

I danced with all of my heart, loving every second of my role as a Flower, and to my complete delight I was offered the chance to join the company on tour and even earn a small stipend. This would be my first adult role in a ballet company, and suddenly my dream of becoming a professional ballerina didn't seem out of reach.

Later I felt a little nervous. It was one thing to be away at a summer intensive with a hundred other kids. There, chaperones watched over us day and night with eagle eyes. It was quite another thing to be on tour with a dozen teens and adults in a professional company. But despite my misgivings about being away from home, I was so proud and honored to be offered this opportunity that I didn't consider refusing.

I knew that as an apprentice with the ABB, I would be

able to share ballet with thousands of people throughout northern New England, and I could show the people in my small corner of the world that black girls could be ballerinas too. Many refugees from the wars in the Congo and immigrants from West Africa had been resettled in Burlington. This would be my chance to reach out to them and encourage their children to dance ballet.

I spent the next several weeks preparing for my tour. I would be missing a lot of school, but my teachers were wonderful, and instead of giving me a packet of work sheets and assignments, they handed me a list and said, "This is the material that you'll need to cover during that time. However you do it is up to you and your parents. We trust you and them to make it happen."

I was getting more and more nervous as my apprenticeship with the ABB approached. "You don't need to do this if you're not comfortable with it," my mother told me, sensing my nervousness.

"But I really want to," I insisted.

My dad said, "You're only twelve years old, Michaela. You'll have plenty of opportunities like this in the future."

He might have been right about that, but I had to take this chance to tour with a professional ballet troupe. It was too great an opportunity to learn. I couldn't pass it up.

When we pulled up in front of Ms. Madeline's house, where I would be living, Mia exclaimed, "Wow! Look at that beautiful haunted mansion!"

Maybe it was the look of fear on my face or maybe Mia

could read my mind, but she quickly added, "Call me on your cell phone if you're afraid of ghosts at night. I'll tell you a story or sing you a song."

As soon as we walked into the old, rambling, three-story Victorian house, we were greeted by an enormous German shepherd. Mia squeaked like a frightened mouse and ran off, abandoning me to my fate.

"Nice boy. Good boy," I said softly, and tentatively reached my hand out to pet the dog, which was larger than me. Finally a tall girl with long, dark blond hair came down the stairs.

"Hi, I'm Lida, and I see you've already met Ms. Madeline's dog, Buddy. He won't hurt you."

Lida was seventeen and would also be an apprentice that year. She showed me my room, which contained two beds. "Ms. Madeline sleeps in a different wing of the house, far from the rooms on this side. There's so much space up here that she said we can have separate rooms if we'd like."

I looked at Lida and noticed that her eyes were as wide and nervous as mine. She added, "Or we can share this room."

"What would you prefer to do?" I asked, hoping that this older girl wouldn't object to sharing a room with a twelve-year-old.

"Share this room," she promptly answered.

I breathed a sigh of relief. "Me too!" I said. We both

smiled at each other, relieved not to sleep alone in the old, creaky house.

She admitted, "I think this house is haunted, so I'm thrilled to have a roommate."

That evening after dinner, my family departed. In the darkness of the November night, the Victorian house looked and sounded haunted. Lida and I scared ourselves into believing that every creak and groan of the old house was really a ghost prowling the halls. I dreaded getting up to go to the bathroom at night, fearful of running into a shadowy specter in the hallway.

Lida would occasionally drive home to Albany on weekends or to a nearby friend's house to stay overnight. On those nights I would have been lonely and scared— terrified, actually—if it weren't for my new best friend, Buddy.

Though Buddy was huge and fierce-looking, I quickly learned that he had a soft heart. Whenever Lida was away, I made him stay with me in my room. Buddy kept me warm and made me feel safe from ghosts, nightmares, and my fear of kidnappers, evil spirits, robbers, and mice. I couldn't help but think that if the Stahlbaums from *The Nutcracker* had owned a dog like Buddy, the rats would not have been able to invade, and the entire plot of *The Nutcracker* would have been different.

•◆•

During my time at Ms. Madeline's, my days were filled with rehearsals. I was cast as a Mirliton (or Marzipan Shepherdess), and performed in "Waltz of the Flowers," the Snow and Chinese dances, and—much to my surprise and pleasure—the Arabian dance! I learned that in a small company, you danced many roles and did whatever was needed. So I became the understudy to the men in the Trepak (or Russian) dance, which caused me to feel like a pronghorn antelope as I leaped high into the air.

The adult dancers at the ABB were all very nice and nurturing to me. They took good care of me, and I felt totally safe with them throughout the entire holiday season. The men were funny. They loved to tease me about my jumping skills and often challenged me to jumping contests. I loved rehearsing the Trepak, and I wished that I would have an opportunity to dance it onstage. But I never did, because none of the men missed a performance.

An important lesson that I learned from this experience with the ABB was that professional ballet dancers didn't have an entire semester to learn one dance routine, as I had in my dance schools. The rehearsal for a ballet lasted only a few weeks, and you had to learn several dances, not just one or two, during that time. This made it much more intense than anything else I had experienced up until then.

The best part of that *Nutcracker* season was the actual touring. We'd travel through the snow by van from small town to small town, to picturesque opera houses decorated with holly and lights.

We worked with a different group of kids in each place, and I never got tired of seeing the happy and excited expressions on their faces when we joined them for rehearsal. To my friends back home I was just another kid, but to these little ballet dancers I was one of the ballet stars, and that made me feel very grown-up.

The adults in our audiences seemed just as happy to see us as the children were. They were always so appreciative. Their applause warmed my heart on those bone-chilling nights.

My favorite role during the tour was as an Arabian. It was exotic and romantic. I was especially happy to dance it for all of my neighbors, classmates, and friends in Burlington.

The visit to the Burlington area was fun for me and the other apprentices. My parents provided us with lots of food, and spread out mats and comforters on our floor so that we could sleep at my house. After we left, I felt a pang of loneliness to be leaving my family again, but we had other performances ahead of us.

As much as I loved the dancing and felt so privileged to be touring with the company, it was still hard to be away from home. One afternoon I got lost on my way back from the pharmacy to Ms. Madeline's house. Another time I was convinced there was a ghost in my room at night. My mom talked me through my anxieties on many long phone calls, and Mia made good on her promise to sing me to sleep on the nights I was really scared.

On Christmas Eve we danced our final performance, in Pittsfield. My entire family came to watch our closing night, and after the show we drove through the dark and snow back to our home in northern Vermont.

Chapter 28

TURNING THIRTEEN

When I returned home from the *Nutcracker* tour, my dance teachers told me that I needed to enroll in a preprofessional ballet program the following school year if I wanted to continue making progress. "Where should I go?" I asked.

There were many good programs available, but none was in Vermont, and we found only two that boarded middle school students. The closest of those was in Montreal—it was three hours away from my home.

My parents drove me up to the school to audition. I was instantly accepted into the ballet school, but I would also need to enroll in a private academic school that boarded ballet students my age. I was disappointed to learn that there were two problems with this plan. All of my academic subjects would be taught in French. And because I was a citizen of the United States, my parents would have had to pay a high fee for my education.

The second school was in Washington, D.C., ten or eleven hours away from home. I imagined myself sick with the flu, waiting for my parents to come pick me up after shoveling ten feet of snow out of our driveway. This was not an option, so while we tried to figure out what to do about next year, I continued to work hard in my local dance classes, and I searched for a summer intensive.

We drove more than one hundred miles over the snowy mountains to Brattleboro, Vermont, where I auditioned for a program. I was accepted, but we learned that it would cost a fortune. I didn't want my parents to pay thousands of dollars.

"Maybe I can get into one of the American Ballet Theatre summer intensives. ABT's having an audition in Boston. Can we go?" I asked.

"Sure," my mother agreed, and she instantly registered me for the audition.

Early on the morning before my thirteenth birthday, we boarded a bus to Massachusetts. When I arrived at the tryouts at the dance studio of the Boston Ballet, I looked at the girls spilled all over the floors, stretching and warming up. I felt a sinking sensation in the pit of my stomach. It looked as if I would be the only black girl auditioning.

I was shaking from nervousness when I entered, but as soon as the class began I focused all of my energy on doing exactly what the dance instructor told us. I decided that I was going to give this my best effort and prove that a black girl could dance as well as a white girl.

Within minutes I discovered that the audition was no different and no harder than my ballet classes at the Rock School had been. For the first time I understood what Ms. Stephanie from the Rock School had meant by the term *muscle memory*, as my body moved automatically into position for the tendus, relevés, ronds de jambe, battements frappés, grands battements, fouettés, assemblés, arabesques, pirouettes en dedans and en dehors, grands jetés, and all of the other steps that I had grown up doing nearly every day of my life.

Toward the end of the audition we switched into pointe shoes. I had planned to take a well-worn and comfortable pair, but in my last class the shank of one shoe had broken. The night before I had hurriedly sewn elastic and ribbons on to a new pair and tried to break them in on the wooden floor of our family room.

Before I went to bed I had squashed them several times by closing my bedroom door on them. For good measure, in the morning I had banged them against the brick fireplace. Now, when I slid my feet into them, they felt just right. As I tied my ribbons I could almost hear Ms. Stephanie reminding me to tie the knots tightly.

Much to my relief, my pointe-shoe ribbons stayed knotted. Nothing went wrong at all in the audition, and I even remembered to smile. As I left the studio, Raymond

Lukens, who was in charge of the audition, winked at me, and my heart soared. Maybe that was a good sign, I told myself. But by the time I had my coat on and was walking out the door, I had managed to convince myself that it was just his way of saying hello and I probably hadn't gotten in.

The staff told us that we'd get the results of our auditions in about two weeks. I didn't know how I could possibly wait that long. Then, a few days later, my mother clicked into her email and shouted, "Michaela! Michaela! Come here! Hurry!"

I raced into the kitchen, and she said, "Look at this email! You've been awarded a full scholarship to the ABT summer intensive!"

"Where?" I asked, because ABT has several locations, the most prestigious of which—in the eyes of young ballerinas—is New York City.

When Mom answered, "New York City," I leaped around the family room, jumping over furniture and two laundry baskets, and nearly knocking myself out by running into the fireplace. But before I could leave for the summer intensive, I had so much more to do. My parents and I had agreed that if I worked really hard at my academics and finished both seventh and eighth grades in one year, I could attend the Rock School's high school program in September . . . if I was accepted. Also I had the 2008 YAGP to finish preparing for.

The YAGP had taken a lot of preparation that year.

My teacher and I couldn't decide what to perform for the contemporary part of the competition, so I had learned three very different dances. Each of these choices demanded something unusual from me.

One was a modern piece taught to me by the modern dance teacher in our studio. It was choreographed to a selection from "Passion" by Peter Gabriel and required strength, flexibility, and agility. At first I thought it would be easy to perform because it didn't require pointe shoes, but it was actually difficult. I felt that I couldn't understand the artistry of this piece. The modern dance felt like a gymnastic routine to me, and because I was so young and inexperienced, I couldn't figure out who or what I was supposed to be in it.

I prefer dances from story ballets that require some acting—ballets in which I can lose myself in the mind and body of another person. I also like routines that contain certain distinctive moves in the choreography. My second contemporary number was a cute and flirty dance en pointe to the song "Partenaire Particulier." It required acting skills and changes of facial expression. I enjoyed stomping en pointe like a pouty young woman, indecisive about her choice of boyfriend.

The third contemporary dance was the opening variation of the wonderful ballet *The Firebird*, choreographed by Mikhail Fokine to the music of Igor Stravinsky. *The Firebird* requires quick and nimble movements, as well as precise ports de bras that imitate the flutter of a bird's wings. I spent a lot of time that year looking at birds and mimicking their movements.

My classical variation, the Princess Aurora role, was the most artistically fulfilling of my dances. I am a romantic at heart, and dancing as an enchanted young princess whose engagement to a handsome prince was about to be announced on her birthday thrilled me. The choreography of the ballet in this variation is stunning! I particularly loved doing the series of ronds de jambe en l'air en pointe, which began about a minute and a quarter into the variation as I started to cross the stage. Nothing made me feel more like a real ballerina than this variation. Its demanding artistry excited me so much.

Then, as the YAGP approached, I decided to experiment that year by trying my hand at choreography rather than worrying about winning. The most exciting part about this was that I convinced Mia to do a duo ensemble piece with me. She was still taking lessons at our local dance school, though they were just for fun and exercise because she was focusing more on her music, taking oboe, English horn, singing, and piano lessons.

"But I don't really dance ballet all that well," Mia protested when I first asked her about it.

"But it will be so much fun! I was thinking we could make it a comedy skit and dance to the music from Johnny Depp's new movie, *Pirates of the Caribbean*. You won't have to dance much. I'll hit you on the head early in the performance, and you can lie on the stage unconscious half the time. The whole idea is to travel to Torrington together and have a great time," I said.

Mia liked that idea, and she agreed to do it. We had a blast planning our costumes. We went to a costume shop in Burlington and bought pirate costumes. Then we went to a party-supply store and found a lightweight pirate's chest and lots of Mardi Gras bling.

We planned the skit so that we danced onto the stage as two sword-wielding pirates dressed in contrasting colors. We discovered a treasure-laden chest and got into a sword fight over it. My pirate knocked Mia's pirate over the head with the flat of her sword and she collapsed onto the ground. My pirate was then so elated about having the treasure to herself that she danced all over the stage, leaping over the chest and Mia's pirate. She got so carried away that she didn't notice that Mia's pirate had revived and carried the chest away, leaving her to celebrate over nothing.

We had so much fun doing this skit. Usually we'd both start laughing so hard in the middle of it that we'd end up rolling around on the floor. Sometimes we'd argue, and each of us would have our pirates doing what we wanted them to do, so we'd often bump into each other and accidentally knock ourselves over. Once Mia got so excited about showing the imaginary audience the bling that she forgot to get up and dance. That really had me giggling until I was breathless. I don't think I can even remember the dance steps in this routine, just the silliness and laughter.

When my mother had told me not to worry about my costumes that year, she meant it. She ordered a professional tutu pattern and made me a feathered firebird tutu in red, orange, and yellow. It was amazing, but even it could not compare to the splendid Princess Aurora costume.

Mom and I could not find the right fabric and lace for the Aurora costume, so we bought a used wedding gown from a local thrift store for only thirty-five dollars! I was filled with anxiety as my mother cut the ivory lace gown into pieces and dipped them in a bath of rose-pink dye. Then I watched with fascination as she sewed the pieces together to make a gorgeous tutu, nearly identical to the ones used by famous ballet companies for act one of the *Sleeping Beauty* ballet. As a final touch she sewed on thousands of tiny crystals. "It's beautiful!" I gasped when she presented me with the finished tutu.

I was so inspired by the combined beauty of the tutu, the choreography, and the music of Tchaikovsky that I felt I danced my personal best at the YAGP regional competition that year. Ever since then I have longed to dance that role in a professional production of the *Sleeping Beauty* ballet.

The YAGP in Torrington, Connecticut, that year was the best experience I ever had at the competition. I won a couple of awards, and ironically I earned a score of 100 percent from one of the judges for the modern dance that I disliked so much. But for me the highlight was the pirate dance. Much to our astonishment, Mia and I earned an overall score of 95 percent! High enough to qualify for the finals in New York City!

After the YAGP I auditioned for the Rock School's high school program and got accepted with a full scholarship for the dance portion of the tuition. I had spent my childhood with the teachers and the kids at the Rock School. I couldn't wait to see them in September, so I signed up to dance at the YAGP in Philadelphia that spring, thinking that I could experiment a bit more with dance.

"You can't win an award in Philadelphia," my mother warned, "because you've already won awards in Connecticut."

"That's okay. It's not about winning an award," I said. I worked on my own to learn a different classical variation and choreograph a new contemporary dance. I didn't want my mother to put any more work into costuming, so I did most of the sewing myself, figuring that if I could sew ribbons on to my pointe shoes, I could sew a couple of easy costumes.

I chose music that lent itself to draped costumes, because using a sewing machine proved to be a lot harder than I thought it would be. I kept worrying that I'd sew my fingers on to the fabric, and I almost did once.

I used a lot of glue on those costumes, but my stitches

and the glue held together long enough for me to perform. Both of my do-it-myself performances scored high enough to get me to New York City, so I chose to take the contemporary from the Philadelphia competition and my Princess Aurora from the Connecticut competition to the YAGP New York finals that year.

While in Philadelphia I had a great time seeing all of my old friends and former teachers. They made me feel I was part of the dance school again. I hugged and cheered a lot the two days that I was there, sitting in the audience with the Rock School kids. And when I went to New York for the finals the next month, I got to hang out with them again.

Chapter 29

AN ABT SUMMER

B efore I knew it, it was summer and time to move to New York for the ABT summer intensive. Mia had decided to study American Sign Language in New York City, so that in the fall she could work with the children in a regional program for the deaf near our home in Vermont. She also enrolled in a piano intensive for teens at Hunter College and one at the Third Street Music School Settlement. It was pure serendipity that the furnished apartment that Mom rented had a piano sitting in the middle of the living room. The piano delighted both Mia and the elderly neighbors next door, who seemed to enjoy listening to her music.

Every morning Mia and Mom walked me to the ABT studios at Twentieth Street and Broadway, and they'd meet me there in the late afternoon to walk me home. We'd discover little shops on the way, and we loved exploring them together.

Despite the attraction of the stores, street fairs, farmers' markets, and restaurants, my life in the city that summer revolved around ballet. I could not afford to lose my focus on the reason I was there.

The competition is fierce for the world-class, top-tier companies. Ballet is probably the only career in which you begin training as a preschooler. Millions of little kids start lessons then, but as they get older and distracted by other things in life, their numbers decrease. Most ballet students drop their lessons in high school because either they want typical social lives, they develop different interests that demand more of their attention, or their bodies have changed. At that point they may come to the conclusion that they just were not born to be professional ballet dancers. If they dance at all, they do it for exercise and pleasure. They usually take up jazz or modern, in which the criteria are more forgiving. But none of that was the case at ABT, where all of the kids were as passionate as I was.

When I stepped into the studio on my first day of class, I was stunned. I had met most of these students at the YAGP. Unlike any other class that I had ever experienced before, this one had only excellent dancers. Each student was there to become even better and make his or her way up the ladder to a professional career in ballet. Though some of us still had some growing to do, my class at the ABT summer intensive was filled with teenagers who were very serious about ballet. They had the bodies, the talent, the competitive spirit, and the drive to become professional ballet dancers.

At the intensive nobody took unnecessary bathroom breaks. No one disrupted the class by clowning around. No one complained. We knew what we were there to do, and we did it. I could smell the competitiveness oozing through my classmates' pores, along with the sweat from their hard work. Yet despite the competitiveness, no one was ever mean to another dancer in the class. No one was a prima ballerina at the age of thirteen or fourteen. The program instilled a sense of corps de ballet in us, and expected us to be civil, courteous, and humble toward each other.

I had learned many tricks in ballet. An example of this would be multiple pirouettes, sometimes eight or more, or arabesques greater than ninety degrees. These steps are not required by any classical ballet choreography in the world, but as kids we were often tempted to show off and outdo each other. I could tell immediately that ABT didn't want us to do tricks. Instead it expected us to develop the appropriate technique for our ages, control over our bodies, and artistry. So I often worked long and hard on some of the easiest steps, such as holding an arabesque at forty-five degrees when I wanted to raise my leg to ninety degrees, or doing a passé so that my toe touched my leg in precisely the right spot. These were all steps that we had learned years before, but that summer we perfected them.

I worked so hard and sweat so much in the warm studios that I wore out a pair of pointe shoes every day. Most brands of pointe shoes are made of paste, cardboard, paper, and satin. Sweat makes them soggy. A pair costs more than fifty dollars, so I tried everything that I could to salvage damaged shoes. I'd line them up on the air conditioner or hang them in front of a fan until I dried the sweat out of them. Then I glued the boxes and shanks to give them extra support. I darned the tips of the boxes. If I was lucky, I could get another half day out of them.

ABT sold us tickets to its Thursday night performances at the affordable rate of twelve dollars a ticket. So when we weren't in dance class, repairing our pointe shoes, or laundering the three leotards and pairs of tights that we drenched with sweat each day, we forged friendships by attending the ballet together at night. Often we'd do sleepovers, giggling into the night but waking up early and heading off together to start another day of hard work.

My friends were from Canada, Japan, England, China, Mexico, and far-flung parts of the United States. When I was younger, I had once asked a ballet teacher how she seemed to know every other ballerina in the world. That summer the answer came to me. The friends whom I made in the summer intensives were from all over the world. As we grew older and joined ballet companies, we would remember the friendships that we forged when we were young. So I felt an awareness and sense of camaraderie that I hadn't known before. I recognized that we would be friends and competitors not only now but also in the future, when we would fill the ranks of the world's ballet companies.

Most of my friends left in late July, but I stayed behind to work for two weeks as an assistant teacher in the ABT Young Dancer Summer Workshop. I had the good fortune to be assigned to Franco De Vita, the principal of the Jacqueline Kennedy Onassis School at ABT. It's difficult to describe how I felt doing that. On a Friday I was a student, one of the kids training at ABT. On Monday morning I was an adult, one of the assistant teachers.

At that point in my life I was one of the youngest children in my family. As such, I had never experienced teaching younger children, and I worried that I might be impatient with them. I was matched up with a class of eight-year-olds, and it came as quite a surprise to me that I loved working with these young dancers. I even went back to teach the next year.

Chapter 30

MY YEAR OF ANGST

Mom, Mia, and I had no sooner gotten home from the most crowded area of New York State and unpacked our bags than we packed up our camping gear and headed off to the most remote part of the state, the Adirondack Mountains. We camped beside a lake, where I spent the last two weeks of the summer swimming, hiking, kayaking, and watching loons with Amie, Mariel, Mia, and our parents. It seemed as if our days on the lake and nights around the campfire, roasting hot dogs and making s'mores, were over too soon.

Almost before I knew it, we were home again and I was rushing around packing for my first semester of high school at the Rock School. On the ride down to Philadelphia my head was spinning. I was in turmoil. I loved ballet with a crazy passion, but I was just beginning to recognize the sacrifices that my family was making on behalf of my future. Shortly before we left I had learned that ninth grade at the Rock School's high school program, as well as my pointe shoes and room and board, would cost my family twice as much as their mortgage that year.

One morning a neighbor stopped by and told my parents that he had just retired. He was leaving for Florida and wanted us to keep an eye on his house. When my dad was leaving for work, I said, "Dad, you're in your sixties. When are you going to retire?"

He laughed.

"Mom, why did you adopt so many children?" I asked.

"Which of you should we have left behind in an orphanage? You?" Mom asked.

That thought sent chills up and down my spine. "Maybe . . . maybe I should stop taking ballet lessons so Dad can retire," I said, unsure of what my mother's answer would be.

"Isn't that your passion . . . your dream?" she asked.

"Of course," I answered.

"Then you focus on your dancing. We'll retire someday, but certainly not now."

As we neared Philadelphia I thought of the sacrifices that I had made too, so that I could become a ballerina. As a child I had loved ballet so much that I had even turned down birthday party invitations so as not to miss classes. I had given up swimming and public school. And now I was giving up the time I could be spending with my family.

I asked myself, *Suppose I wake up someday and decide that I hate ballet? Will I regret the sacrifices that I have made?* As hard as I tried to imagine such a day, I couldn't. I felt that ballet was in my bones and in my blood. It was all wound up with who I was. I would give up breathing air before I would give up ballet.

When we pulled up in front of the Rock School, I was full of optimism and floating on a cloud. I was certain that it would be like living in paradise and I'd have a perfect year. But as soon as my parents helped me unload and said their goodbyes, my feet touched the ground. By the next morning I was miserable. We had been assigned roommates by age, so my close friend Ashley had gotten a different roommate. Though my roommate and I would eventually become friends, at that time we didn't know each other. Whom could I talk to late at night when I was feeling insecure?

To make matters worse, one of the adults in our dorm reminded me of Auntie Fatmata, so that's how I referred to her in my mind. I was convinced that she hated me. This made me homesick, and I felt like Number Twenty-Seven again.

It took me a couple of months to learn that most of the other girls thought that "Auntie Fatmata" hated them too. She would give us detention and ground us for a month for even slight infractions. I was thirteen and rather rebellious at that age, so there were times when I definitely broke rules. I expected to be punished for that, but as soon as I learned the rules, Auntie Fatmata made new ones, which she'd often fail to tell us about until we broke them. I remember getting a month of grounding for eating a granola bar in the lounge and another month of grounding for sharing my muffin with another student. To me the only good thing about those groundings was that we had to spend them studying, so I was way ahead in my schoolwork, and my grades were great as a result of all that punishment.

I was one of the youngest kids in the high school program. Except for my early years in Africa, I had been sheltered by my parents throughout my life. Now I was around older teens day and night. Some of them took me under their wings. What they taught me would have made my parents' hair stand on end, but at that time I really believed that these older kids knew everything.

That year I learned from an older student that alcohol mixed with a power drink would relax my muscles, relieve the stress of Auntie Fatmata, and ease the pain of tendinitis. Someone suggested I try it once when I was off campus. I did and never tried it again because it made me violently ill.

One of the older girls recommended an all-tea diet to

lose weight. I tried that for one day, and shook so badly that I could barely stand en pointe. I thought I had some terrible disease and called my mother for medical advice.

Someone else told me that smoking cigarettes would help me relax as well, and they had the added benefit of helping me lose weight. But, like tea, cigarettes made me shake. They also gave me a sore throat and asthma, neither of which helped me relax.

One girl recommended skipping meals, taking laxatives, and vomiting after meals in order to keep thin. I'd often watch her cut her food into tiny pieces and spread it all over her plate to make it look as if she had eaten. When I was little, I had experienced starvation and dysentery. I knew what it was like to starve or vomit until you were nearly dead, and I decided not to control my weight in any of those ways.

My father worked in New Jersey, so he visited me once a week. On one of his visits he sighed and said, "Michaela, I know you'll find this hard to believe, but there's a reason for the minimum age for smoking and drinking." I felt so contrite when I saw the disappointment in his eyes.

My favorite saying had always been "To thine own self be true." Being true to myself meant throwing myself into my ballet, and not letting anything get in the way of my goals.

I had been placed in the highest-level ballet class at the Rock School. I felt that I needed to prove to myself that I really belonged there. The auditions for the Rock School's

Nutcracker were coming up. I worked hard preparing for them. I wanted to dance flawlessly during the auditions so that I would get a choice role.

"Don't be disappointed if that doesn't happen," Mia warned me during one of my telephone calls home. "You're only a freshman. The Sugar Plums will probably fall to the older girls."

On the day of the auditions it was clear to me that older and taller girls would be cast as the Snow Queen, the Sugar Plum Fairy, and the Arabian woman. Children in classes lower in level than mine would be cast as Clara (the character also known as Marie in many productions), Party Children, Soldiers, and Pollys. I suspected that I'd be a Flower in "Waltz of the Flowers" or a Snowflake in the snow scene. *Smile. Look happy. Dance your very best*, I told myself. *You'll have more opportunities in the future.*

When the cast list was posted, I saw that I had been given the role of Dewdrop. I was most familiar with the Balanchine *Nutcracker*, from having watched the video practically every day with Mia as a kid. In that production Dewdrop is the lead character in "Waltz of the Flowers," and she tiptoes among the flowers as she dances her solo. Since she is only a tiny dewdrop fairy on the petals of the flowers, she is supposed to be small and delicate. My greatest desire as a ballerina was to be delicate, so I was very grateful to be given this dainty but challenging role.

When rehearsals began, I was shocked to discover that in the Rock School's *Nutcracker*, Dewdrop dances among the other characters throughout most of the ballet and has two rather long solos. It was going to be intense to learn all of Dewdrop's moves with two separate casts, while at the same time learning a classical variation from *La Esmeralda* and a contemporary dance for the 2009 Philadelphia YAGP in early January. But I loved a challenge.

My parents drove down from Vermont to see me perform in *The Nutcracker*. My mom later told me, "As we waited in the audience during intermission, Ms. Stephanie came rushing over to tell us that you were injured. You might not dance the second act. We were crushed, but we said that with the YAGP coming up, you needed to care for yourself."

Even Ms. Stephanie didn't know that I would dance until I entered the stage smiling. My ankle was painful and tender, but I didn't let it show on my face. I would never allow the audience to see how I was feeling. Even at thirteen I believed that the audience should think that the hardest combination of steps was effortless, and a ballerina's personal trials shouldn't show on her face. So I was overjoyed to dance the second act flawlessly, looking as though I didn't have a care in the world. After the show I was able to moan as I iced my ankle and then hobbled off to greet my parents.

I recovered at home on my semester break and returned two weeks later, ready to compete in the YAGP. I had always wanted to dance *La Esmeralda* in competition, even though this variation, danced with a tambourine in hand, begs for criticism. Ms. Natasha, who was my competition coach once more, warned, "The tambourine is not just a prop. It's a part of your dance. If you don't use it artistically, the judges will tear you apart!"

Of all the classical dances approved for use in the YAGP, *La Esmeralda* might be the most difficult, because of the staccato beat of the music. When I was dancing this variation, it was tempting to move my body stiffly to accompany the staccato, but this would have made me look jerky, like a marionette.

Esmeralda is a poor Gypsy girl who is in love. When I danced as her, I focused on making my body move smoothly. I made an effort to flirt with the audience so my character would seem more relatable and less mechanical.

On the day of the competition I did my best. I dropped my heel during a pirouette, and though I danced for the crowd and the audience loved it, I wasn't convinced that the judges loved it. That's why I was totally blown away when I was awarded the Youth Grand Prix, the top award for my age division.

··➤· *Chapter 31* ·➤·

FIRST POSITION

I n the fall of my freshman year of high school, I received a letter from ABT informing me that I had been
named a National Training Scholar. This meant that I had
received a full scholarship to the 2009 ABT summer intensive in New York City. This scholarship covered both
my tuition and my board. That year the summer intensive students would be staying at a college dormitory, so
I spent the summer of 2009, when I was fourteen, in the
city without my mother and sister.

While I was away, my family moved into a house across
the street from the high school that my sisters would attend. It was a larger house, and their timing was perfect.
Though Amie, now an adult, had moved out on her own,
our family grew by two more children that summer.

Bernice and Jestina, both nine years old, were the adopted daughters of my brother Adam and his wife, Melissa. Before their adoption in 2003, these two little girls had been living in a Liberian orphanage. It was one of the poorest and most unsupervised in the world. Chaos reigned in their orphanage of 295 children and a single caregiver.

They had come to Adam and Melissa almost like feral children.

With no other experience raising children and no formal training in child development, Adam and Melissa were at their wits' end with the girls. When Adam and Melissa separated, caring for Bernice and Jestina proved to be overwhelming, so my parents accepted guardianship of the girls.

When I came home from the ABT summer intensive, I struggled to catch up with American Sign Language, or ASL, which the family used with Bernice, who was deaf. Unlike Mia and Mariel, I had no talent for ASL, but Bernice and I bonded when I taught her ballet, hooking up the stereo on the deck so that the wooden floor reverberated with the beats. Bernice had such a natural ability for ballet that it nearly broke my heart that her deafness prevented her from studying it professionally.

One of the problems of teaching dancing to deaf students is that the music needs to be played very loud in order for them to feel it. Few dance schools can do that. The volume could injure the other students and teachers.

I would have loved to teach her all year round, but the summer was drawing to a close and I'd soon have to return to Philadelphia. However, my experience with Bernice convinced me to include deaf students when I owned my own studio someday.

Over the summer I learned that "Auntie Fatmata" wasn't returning. I was vastly relieved about that but still torn about going back to Philadelphia. Living apart from my family was painful to me. I would never have done it for anything other than ballet.

I was thinking about all of this when my mother popped her head into the doorway of my room. She told me that she had received an email from a woman named Bess Kargman, who was producing a ballet documentary.

"She asked if you'd want to participate," my mother said.

"Do you trust her?" I asked.

"Yes, she comes recommended by the YAGP. She received approval from them to follow some of the kids who competed, if the parents give their permission," Mom explained.

Thinking that it involved only a brief interview, I agreed. Once I was back at school, Bess began to come around with her camera and crew. At that time I was very shy about being videotaped, and Bess wanted to videotape me doing everything. She would follow me around, taping my meals, my dance classes, and my rehearsals. She followed me home to interview my parents and siblings.

She even followed me backstage during competitions. Sometimes when she'd come, I'd hide in my closet, hoping that she'd leave if she couldn't find me.

My mother said, "Michaela, you've always talked about wanting to make the world aware that black girls can be ballerinas too. Maybe Bess's documentary will help you accomplish that."

I gave some serious thought to that and decided to make myself more available to Bess. Even with a purpose in mind, though, I still found it difficult to participate in the filming of *First Position*.

In the orphanage I had learned to be stoic and hide my true feelings, so I was not used to expressing myself openly. When Bess arrived, she wanted me to share my feelings. She could see right through insincerity. She asked hard questions and demanded honesty. This was especially true when she asked me about my wartime experiences in Sierra Leone and my life in the orphanage.

The memories I had of my early years were painful ones. Just thinking of the debils and the dead bodies that littered the streets of my homeland gave me nightmares. Many times I was reduced to tears during the filming. After an interview I would have difficulty sleeping for days.

During one interview I told her my birth name. When I realized what I had done, I became very upset. I worried about that for months afterward. Because Bangura is the most common name in Sierra Leone, I worried that some unrelated person with that name would claim to be my parent and take me away from my American family.

Bess's prying camera caught me during moments of joy, like when I greeted my parents after my successful performance in the 2010 YAGP in Philadelphia, and times of contentment, like when I picnicked with my family on the deck of our home on an unseasonably warm spring day. It caught me during moments of despair, as I suffered from a painful case of tendinitis during the YAGP New York finals. Before I stepped onto a stage, I needed time to center myself and concentrate on what I was about to do, but Bess was right there with her camera. *Remember, you can reach out to people and let them know that black girls can be ballerinas too*, I'd ultimately remind myself whenever I felt uncomfortable under the eye of the camera.

While I was emoting for Bess, I was also looking inward and beginning to analyze who I was. I decided that I didn't particularly like the grumpy, pushy, selfish girl who would sometimes appear, especially because those aspects of my personality interfered with my dancing. I was striving for a dance style that was strong yet soft, delicate, and gracious. When Bess was filming me, I was fourteen and already able to dance with excellent technique, great strength, and agility, but the gentle, soft, delicate, and gracious part was still missing.

I knew that I needed to change the way I acted if I wanted to achieve my goals in ballet, but I had no idea how to bring all that together. In the Rock School studios Ms. Stephanie struggled patiently and endlessly with me to bring out the soft side of my dancing and personality, but I felt that I still had not achieved enough of it by the time I was fifteen years old.

At the 2010 YAGP I was awarded a full year-round scholarship to the Jacqueline Kennedy Onassis School, or JKO, at ABT. Bess recorded that moment in *First Position*, and I am glowing in that scene. This scholarship would cover everything: room, board. . . . It lifted a heavy burden from my family.

Yet when people discuss *First Position* with me, I learn from them that it isn't the happy glow of my face that they remember.

I had been distraught when Bess videotaped me sitting with my foot packed in ice shortly before I had to perform in a dance competition, and it is this image that many people take away with them. This image lets the world know that injuries and pain are facts of life for us. It symbolizes all the sacrifices that ballerinas make for their art.

Many people who saw Bess's documentary ask me how I was injured and why I continued to dance that day. I tell them that those beautiful satin pointe shoes that I had longed for since age four are not sturdy footwear. It's common for a ballerina to come down from a leap, or out of a turn, and twist an ankle or pull a tendon. My weakest spot was my Achilles tendon. In *First Position* I faced the possibility of being eliminated from the YAGP because of my injury. The decision was mine—totally. I asked myself, *Should I dance or should I sit it out?*

My teachers recommended that I sit it out. They were concerned about any long-term effects of dancing on an injured foot. I was only concerned with winning a scholarship to a school that every dancer aspired to attend—a school that would guide me to my ultimate dream.

That day I squeezed my swollen foot into my pointe shoe and pretended that my tendon didn't hurt. I danced onto the stage and did a variation that required many grands jetés. When people ask me how I did that, I honestly don't know. I think that when I stepped onto the stage and heard the roar of the audience, I had such a rush of adrenaline and joy that I didn't notice the pain.

Chapter 32

GROWING UP

Sometime between the ages of fifteen and eighteen, I grew up, both as a human being and as a dancer. It didn't happen all at once. I attended the ABT programs, both the summer intensive and the year-round program at JKO. The first year was bumpy for me. It had its ups and downs, emotionally, physically, and artistically. I found it so difficult to express the artistry and emotion that I needed for my roles. This frustrated not only me, but some of my teachers as well. Sometimes I wanted to burst into tears in class because I sensed their disappointment.

I was always a very down-to-earth person. Acting came hard to me, and ballerinas need to be actresses as well as dancers onstage. But as I grew from a young teen to a young woman, I finally gained the artistry that had been lacking in my dancing. I also acquired the graciousness that had eluded me at thirteen, and I made it a part of my dancing and my daily life.

When I was younger I had tried to act grown-up by hiding my thoughts, feelings, and behavior from my parents. I had thought that this was a sign of adulthood. Ironically, I discovered that as I grew older, I became more open with my parents. At the age of thirteen I went to my friends for advice, but the advice that they gave me usually wasn't very good. But at the age of eighteen I went to my parents. I found that they gave much better advice, especially about health, money, and many of the other complicated issues related to the world of adults that I was then facing.

Fortunately, my family was able to move to New York City when I was sixteen. When I turned seventeen and got a job in a professional ballet company in New York, I finally could have moved into an apartment with a friend. But I surprised myself by choosing to continue to live with my family for another year.

I've often heard it said that professional ballerinas neglect their education in order to achieve their goals as dancers. There was no way I would do that. My family values education too much. My father would often say, "Suppose a taxicab runs you over tomorrow, and you can never dance again. What would you do without an education?"

Such a gruesome possibility! I didn't want to think of

something like that happening, but my dad forced me to, so I worked hard in school. I attended an excellent accredited online high school for four years. Often, friends who attended private or public schools would comment on how lucky I was . . . how easy it was for me to do my schoolwork whenever it was convenient for me. I laughed at that. Every course in my school came with a thick, heavy textbook. Unlike regular high schools, where the teachers often skip over some pages or chapters, my high school required its students to read every word on every page of every textbook, and the tests were designed to make sure that we did. Every section of every chapter was accompanied by a quiz and essay questions.

How I dreaded those essay questions! I wrote at least two hundred essays every semester. I hated those essays so much that I even took precalculus as an elective, just to avoid them. Then I discovered that my precalculus course included a major research paper that required tons of writing. When I finally graduated with honors in 2012, I felt proud of that accomplishment, and relieved that I no longer had to face those five-paragraph essays late at night after dancing all day.

I was a typical kid in many ways. When I was thirteen, I spent most of my winter holiday crying about a boy who said he liked me one day and would change his mind the next. Once I asked my mother, "How will I know if a boy loves me?"

She said, "He'll be your friend. He'll make you happy.

He'll give you space and remain faithful to you. He'll respect your choices. He'll let you soar and not try to clip your wings." Then she concluded with a smile, "And he wouldn't dream of making you cry through the holidays."

When I was almost seventeen, despite my many hours in the dance studio and my nights at home with my schoolbooks, I managed to find time to fall in love. I was lucky enough to fall in love with a young man who was capable of doing all the things my mother had described to me. His name is Skyler; he's a ballet dancer and choreographer. Skyler understands how important ballet is to me and remains faithful to me when I travel the world. We share dreams for the future. One of them is to someday dance with the same company and be cast as partners in pas de deux.

It was fortunate for me that I had matured emotionally when I did. Without that newfound maturity I probably could not have handled all of the changes that took place in my life after *First Position* was released in September 2011.

·•· *Chapter 33* ·•·

AFTER *FIRST POSITION*

I n August 2011, *First Position* was accepted into the Toronto Film Festival. Bess arranged for Mia and me to see
it before its premiere at the festival. Tears poured down
my face as I watched it for the first time. I wasn't crying over the telling of my story or my memories of pain-
fully dancing through my injury or of being awarded my
scholarship. I was crying because I was overwhelmed by
emotion for what I sensed the film had unleashed. I knew
that my life was about to change in a way that I could not
predict. And that scared me. I was a very private person. I
had opened up to Bess, but I knew that now others would
expect this of me too.

In addition to the Toronto Film Festival, Bess was invited to enter *First Position* in other film festivals, and

her film was nominated for several prestigious awards, including the NAACP Image Award. It gained the attention of the whole world. Soon everyone was watching it, including producers of television programs, the news media, and artistic directors of ballet companies.

As the film's fame flourished, interest in me as a person and a ballet dancer grew.

Suddenly newspapers, magazines, and television programs contacted me. I wondered what I should do. I could either hide from them or accept their invitations. I had always wanted to be a role model to little girls and an activist for change. Here was my opportunity.

As a result of *First Position*, I had the good fortune to be featured in quality magazines like *Marie Claire* and *Teen Vogue*. Next, invitations from television programs came pouring in. I was interviewed for ABC's *Good Morning America* and *Nightline*, and in April 2012 I was invited to be the AT&T Spotlight guest on *Dancing with the Stars* and was flown to Los Angeles with my mom. Soon I managed to control my shyness and enjoy these wonderful experiences.

As exciting as some of these moments were, my first focus was still my dancing, so to me the most wonderful outcome of the attention I received from *First Position* was an invitation by Dirk Badenhorst, the CEO of the South African Mzansi Ballet.

Mr. Badenhorst had seen *First Position* and requested permission to watch me during one of my classes at JKO. Initially, all I knew of his plan was what the JKO principal told me: "Dirk Badenhorst, from South Africa, will be in the class to observe you."

When I learned after class that Mr. Badenhorst wanted me to dance a principal role in *Le Corsaire* as a guest of his company, you could have knocked me over with a feather. I immediately telephoned my mother from the dressing room, but I was so excited I didn't realize that she couldn't make any sense of what I was saying. When I finally got home and explained what had happened, my family was shocked that I was going to South Africa on my own.

Later that night my mother came to me with a bulging file folder. It was labeled AFRICAN ADOPTION. She said, "I was planning to give these to you when you turned eighteen, but if you're going to Africa I think that you should be aware of what is in them." Then she showed me a series of articles about a group of children from Sierra Leone whose alleged parents had come forward, claiming that their children had been stolen from them during the war. They said that their children were trafficked to the United States and sold to white Americans.

Mia, Mariel, and I were on the list of children. I began to protest. "This isn't true," I insisted. "This man who's claiming to be Mia's father . . . he can't be. She told me every night that she had seen her father run down by a truck full of laughing debils." Then I began to tremble, fearing that I'd find that Uncle Abdullah was trying to reclaim me.

I continued reading, and when I saw that a woman with the last name of Bangura was listed as my mother,

I became indignant. "The name of this woman who is claiming to be my mother . . . well, that's not my mother's name. I had my father's last name, but my mother's last name was different. And I saw my mother's dead body, I . . ." I began to cry, not as much from sorrow as from anger and frustration that people should come forward now, when we were nearly full-grown, and insist that we return to a world we didn't know.

"Yes, you're probably correct about your parents," my mother agreed. "Papa Andrew told me that both of your parents had died, and you had remained with your mother till the end."

Then she asked all three of us, "Do you want to pursue this further? Dad and I are willing to help you if you want to contact these people who claim to be your parents. I want you to give it a lot of thought. Here, take this file, then talk about it among yourselves and sleep on it."

Even though the woman who claimed Mariel really was her biological mother, Mariel refused to discuss it with us that night. "I'm perfectly happy being Mariel DePrince," she announced, and then she rolled over and fell asleep.

Mia and I stayed awake late, talking about Sierra Leone and the articles Mom had given us. We realized that once we were out of Africa and in the heart of a loving family, that place and all of the sorrow we had experienced had eventually ceased to exist in our minds. However, since the release of *First Position*, the media had brought it up so frequently that we had to address it.

There were many articles that criticized the way our adoptions had been handled. The adoptions were described as human trafficking, and worse. In a blog an American activist said that we might have been better off with our birth parents. This angered me. We had hidden behind trees to avoid being shot by debils. We had arrived here so sick that it's likely we would have died if we remained in Africa much longer. I needed abdominal surgery within weeks of my arrival. Mia and I truly believed that our adoptions had been lifesaving.

What did bring tears to our eyes that night were the notes our mother had taken when we told her about our lives in Africa as little orphan pikins.

"This is so sad," Mia cried loudly.

"Oh, these poor little girls. What a sad story!" I said, wiping my tears and sniffing.

Mia's eyes grew round, and she laughed through her tears. "Michaela, *we* are these little girls!"

I laughed and cried at the same time. "Weren't we so pathetic back then? Our lives were so hard; we were so sick all the time. But I barely remember this. When I think of being a little girl, I remember Daddy taking us to *The Nutcracker* and Mom baking holiday cookies with us."

"And campfires. And Teddy throwing us into the air and swinging us around the playroom," Mia said.

"And taking us to the movies and the park, and trick-or-treating," I added.

"Remember the time . . ."

Before we knew it, we were reminiscing about Teddy and our childhood in America. As we laughed and cried over memories that had accumulated over thirteen of the seventeen years of our lives, I realized that we had gotten off track. "Mia, we're supposed to be talking about reconnecting with Sierra Leone; instead we're talking about our family."

Mia opened her eyes wide again. "Okay, did you hear what you just said?" she asked.

I looked at her with a sheepish grin. "Yeah, *our family* . . . I got it."

When we returned to the topic of Sierra Leone, our discussion centered on what we could do for the people there, especially the women and children. The boys and girls, but especially the girls, lacked opportunities for education. Education wasn't free, and families couldn't afford the hefty school fees. Three-quarters of the women couldn't read and write. One out of eight women died in childbirth. More than 90 percent of girls in Sierra Leone endured genital mutilation. Recently, laws were passed criminalizing rape and domestic violence, but the country lacked the funds to enforce the laws.

"So what will your answer be?" Mia asked. "Mom asked you if you want to pursue contact with the people claiming to be your family members."

"No, I'm not interested in finding biological relatives. I'm Michaela DePrince now, and that's who I've been for a long time. But when I'm older, I'd like to start a free arts school in Sierra Leone and teach ballet there."

"I can help you. I'll bring over a pile of instruments and teach music," Mia offered. "But I'm not ready to return now . . . maybe in twenty years."

"Me too," I murmured. "In the meantime we can figure out a way to raise money for education in Sierra Leone."

Even though I had come to terms with who I was and the role Africa had played in my life, on the day before my departure for South Africa, I awoke in the predawn hours, sweating and shivering with fear. My dreams had been filled with dreadful memories of Africa. It was the first time in years that the terrors of my early childhood had come back so vividly to haunt me. This made me wonder if I was ready to return.

More than anything else I feared being kidnapped and returned to the home of Uncle Abdullah. I got out my computer and did a Web search. I discovered that it was a five-hour flight from Johannesburg, South Africa, to Freetown, Sierra Leone. I pulled out the huge atlas that our oldest brother had given to us. I was reassured to see that the Kenema District, where I had been born, was two-thirds of a continent away from Johannesburg, a distance of over three thousand miles.

When I shared my worries with Mia, she said, "I'd estimate that it would take at least ten days for someone to drive that distance. I don't think it's worth anybody's time or trouble to kidnap you."

··· *Chapter 34* ···

RETURNING TO AFRICA

Despite my sister's reassurances, I was jittery and nervous at four-thirty in the morning, the time I left for the airport. I hadn't even gotten on my plane yet when I called my mother to tell her that I missed her.

"Already!" she laughed. Then her voice grew serious, and she asked, "Are you feeling okay?"

"I'm just a little scared, you know . . . of going to Africa."

"You'll be fine. Dirk Badenhorst told me that he and his company would take good care of you."

I usually fall asleep easily in any vehicle with a humming motor, but I was too nervous and afraid to sleep at all during the twenty-hour trip to South Africa. My knees shook as we took off, and they continued to shake until I saw the smiling face of Mr. Badenhorst when we landed.

I could not believe how welcomed Mr. Badenhorst's ballet company made me feel. It was a lovely classical company with caring and friendly directors and dancers. I would have loved to work with them full-time, but the trip to South Africa from my parents' home was far too long. My parents were in their sixties at the time, and I knew that the trip to visit me there would be not only expensive but physically hard on them too. Besides, I loved my adopted country, and I longed to dance where I could return to it frequently.

There was a group of dancers from Cuba there too, and they tried to convince me to come to Cuba and dance with the Cuban National Ballet. *Cuba isn't so far from the United States*, I decided, but then I remembered the political situation and realized that despite the allure of that beautiful company, joining it would be impossible for me. So I had to enjoy the experience of performing with these wonderful dancers while we were all far from our respective homes.

Until now I had only danced in two full-scale ballets. One was *The Nutcracker*, of course, and the other was *Abdallah and the Gazelle of Basra*, in which I had danced as a guest of De Dutch Don't Dance Division in The Hague, Netherlands, while a JKO student.

The South African Mzansi Ballet's production of *Le Corsaire* would be my debut performance as a professional ballerina. I had always expected that my debut as a professional would be as part of a corps, so I was bedazzled to be dancing the role of the slave girl Gulnare, partnered with Andile Ndlovu, a South African dancer.

I had wanted to spend all of my time in South Africa perfecting my role. I had danced some of the variations of the other female lead, Medora, but I had never danced the Gulnare variations and the pas de deux. I felt that it would require all of my attention, but I quickly learned that in South Africa, ballet was big news and ballerinas were almost like rock stars! So immediately I was whisked off to a television interview.

My time in South Africa was filled with frenetic energy and frantic activity. I was scheduled for interviews, speaking engagements, lots of rehearsals, costume fittings, and of course the performances. I had arrived on the seventh of July, and *Le Corsaire* opened twelve days later! I had never learned anything that quickly before! I loved every second of it, though I must admit that at times I became very tired. During one televised interview I felt my eyelids drooping and I prayed that I wouldn't fall asleep in my chair.

Before I left for South Africa, my father had warned me that Johannesburg could get very cold in the winter. When I asked Andile about that, he said, "Oh no, the winters are very mild, even in Joburg."

On the day I departed, my mother tried to shove extra sweat suits into my luggage, and I had protested, worried that I wouldn't be able to lift the bag. During my visit, Joburg experienced a cold snap and one of its rare snowfalls. This delighted the residents of South Africa, but I froze.

The weather might have been too cold. The schedule might have been too crazy. But the experience of the ballet was worth everything. Mundane worries that I had during rehearsals, like being dropped during a high lift or twisting an ankle in a turn or, horror of horrors, forgetting the choreography—these all disappeared when I stepped from behind the curtain and into the life of Gulnare.

My entire experience there was enlightening. The prima ballerina in the company taught me by example that it was possible to be on top and still remain genuine and generous. I learned from the very kind director that there was no need for someone in his position to remain aloof and haughty as others often do. And the children of South Africa taught me something about myself.

The U.S. embassy had invited me to give a motivational talk in a school. At first I was scared to do this. I was remembering my middle school years and all of the chatting that went on in the classrooms. I don't have a strong voice, and I worried that I wouldn't be heard over the noise of the students' voices.

When I entered the school and was escorted to the class where I would give my talk, my knees knocked together. *This is really traumatic,* I thought as I took a deep breath and opened my mouth to introduce myself. Suddenly silence fell on the room. Not a voice could be heard other than mine. All eyes were upon me.

The rapt attention of the students gave me courage, and I began to tell my story. As I spoke to them, much to my surprise, I discovered that I was actually better at this than I had thought I would be. These courteous and dignified students probably motivated me far more than I motivated them that day. With my confidence buoyed, I realized that a ballerina could do more than dance.

After *First Position* was released in 2011, people on the streets began to notice me. They'd come up and ask, "Are you Michaela DePrince?" At first I dreaded this attention. I felt too shy to respond with anything other than a quiet, mouselike hello. After I returned from speaking to those kids in South Africa, I found my voice. Now I look at every encounter as a means of touching a life.

·✦· *Chapter 35* ·✦·

FINDING A COMPANY

The trip to South Africa had revived my flagging spirits. In the spring my self-esteem had taken a battering. I had worked so hard that year. Not only had I danced from morning till night both in Level 7 at JKO and with the ABT Studio Company, but I had completed high school while doing so. Then Franco De Vita told me that I would not be dancing in the studio company the following year. He said that I was ready to audition for a professional company. That day when I arrived home, late as usual and exhausted from many hours of dancing, I burst into tears. My heart was broken, my dreams dashed.

I had expected to spend another year in training with the studio company. But here I was, cut loose at seventeen, my dream of joining ABT's professional company shattered. Instead of languishing in my misery, I picked up my broken heart and began auditioning for North American classical ballet companies.

One company told me that at five feet four and a half inches tall, I was too short, and eliminated me immediately. I traveled over a thousand miles to another audition, only to be refused admission. That company told me that it hadn't received my resume and head shot, yet I had confirmed its receipt of both before flying there. I auditioned for other companies, and was always one of five or six dancers remaining at the end of each audition. Yet I wasn't hired by any of them.

Of course, I had known since I was eight years old that classical ballet companies were predominantly white. That probably isn't even the most accurate way to describe the major classical ballet companies. Now, in my anger and frustration, I could find no nice words to describe their lack of black female dancers.

Nine years before, I had searched eagerly for a black female face among the white ones and hoped that, with time, attitudes would change and more would appear. When I looked again, there were fewer black ballerinas. Lauren Anderson had retired from the Houston Ballet, and a few other black women had given up their battles of trying to make it in classical companies. No one had replaced them.

During those dark days, I met Alonzo King when his

famous contemporary company, Alonzo King LINES Ballet, performed in New York City. He invited me to take a company class with LINES, and afterward he asked me about my dream for the future.

I admitted to him that my dream had always been to dance with a classical company. He understood this and respected it, but he also told me that if I someday grew tired of classical ballet, I was welcome to join his company. I was honored as much by the fact that he understood that I needed to be true to myself as by his invitation.

It came as no surprise to me that the only companies that welcomed me were Alonzo King's LINES and the Dance Theatre of Harlem's new professional company. Both were predominantly ethnic, and neither was a classical company. They were both wonderful companies. I would feel proud to dance with them, but I knew that my heart wouldn't belong to either of them, at least not now, when I longed to dance with a classical company.

The Dance Theatre of Harlem, or DTH, was a neoclassical company, heavily influenced by George Balanchine. Arthur Mitchell, the first black dancer to perform with the New York City Ballet, and Karel Shook, a white dancer and ballet master, believed that there should be a place for black dancers in the art of ballet. So they founded DTH.

From 1969 until 2004, DTH traveled the world, performing to great acclaim. Then, under the pressure of tremendous debt, it closed its doors, keeping only its school and ensemble open in order to provide ballet training to the black children of Harlem and beyond, preparing them for careers that, sadly, were almost nonexistent.

DTH had had forty dancers, and it was able to perform full-scale ballets in its heyday. Now, after heroic fund-raising by its board of directors and several major corporate donations, DTH's professional company was about to return, but with only eighteen dancers.

I auditioned for it. I was accepted, and I was most grateful for the opportunity extended to me by Virginia Johnson, its former principal dancer and now its new artistic director. I wanted to contribute to the company's return to the world of professional ballet. I wanted to be a part of its return to former glory. However, I knew deep down inside that my heart belonged elsewhere.

DTH sent me a letter of intent that needed to be signed by the twenty-sixth of April. Once I signed it, I'd be committed to DTH for the 2012–2013 ballet season. I agonized over that decision. Because of my experience with Arthur Mitchell, I felt affection for DTH, but was that enough? As I stared at that letter of intent, I felt like Juliet, agreeing to marry her cousin Paris while longing for Romeo. DTH, like Paris, deserved more than affection. It deserved the love that I couldn't give it.

I decided to spend a year with it. The experience would be invaluable to me, and hopefully some of the media attention I was getting would benefit DTH. If, at the end of that year, my soul wasn't satisfied by the company's repertoire and I didn't feel passionate about it, then I could audition elsewhere.

If the United States didn't want me, perhaps Europe would, I told myself. I knew that I needed to be eighteen to qualify for a European Union work visa. I decided that once I turned eighteen, I could seek work with the classical companies of Europe.

Finally I signed the DTH letter of intent and faxed it exactly on the deadline. A few days later Franco De Vita invited me to officially join the ABT Studio Company, and two months later ABT offered me the position of company apprentice. As an apprentice I would learn its repertoire and perform in its production of *The Nutcracker.* I would then enter the company as a first-year corps member during its Metropolitan season in the spring of 2013. I was speechless upon hearing this news. Here was my dream, tantalizing me like a chocolate ice cream cone, and I had to watch it melt.

I suppose I could have tried to get out of my commitment to DTH, but my parents had raised me with integrity, and I knew it wasn't the right thing to do. I decided to put my dreams of a classical company away for a while and audition with ABT again the following year.

I threw my heart and energy into DTH, determined to do my best during its 2012–2013 ballet season. I made friends among the dancers, and I learned so much from an artistic staff that I greatly admired.

While with the company, I had opportunities to visit

places in the world that I never thought I would see. We traveled to Turkey and Israel.

In Tel Aviv, I swam in the Mediterranean Sea for the first time in my life. Then we traveled across the country to Jerusalem. I felt a special link with my mother there. When she was in the third grade, her pen pal was from Israel. She had always longed to visit the country, but she never had the opportunity to go.

I felt as if I was going in my mother's place. I even left a prayer for her in the chinks of the Wailing Wall, and I wore my *hamesh* (or *hamsa*), a hand-shaped charm, for protection during our travels to the Dome of the Rock and the salty Dead Sea.

Mom had insisted that I wear the *hamesh* she had given me. Muslims believe that it represents the hand of Fatima, the daughter of Mohammed, and Jews believe that it represents the hand of Miriam, the sister of Moses.

Mom explained to me that thousands of years ago, when the pharaoh was killing Jewish baby boys, Miriam had watched over her baby brother, Moses, after their mother floated him down the Nile River to protect him from the pharaoh's wrath. He was then found by the pharaoh's daughter and raised as a son of Egypt. Mom believed that when I wore the *hamesh*, I would be protected, just as Moses had been. I simply believe that the *hamesh* represents my mother's hand, reaching out to protect me from half a world away.

Besides traveling to exotic lands, I had other wonderful experiences with DTH. I was cast in the role of the Black Swan in an excerpt from *Swan Lake*, and I relished the chance to demonstrate my technique and grace as a classical dancer. This was a tough role, and I struggled with the artistry of it. It wasn't enough to perform a technically perfect dance. I needed to *become* the Black Swan, not just dance. I needed to seduce my partner, not merely flirt with him.

One review shook me up. The reviewer said that the Black Swan was trying to woo Prince Siegfried with her rock-solid technique.

"I think that's a wonderful review!" one of my friends at DTH told me. But I knew that it was a terrible review. I told myself that if I wanted to make it as a classical ballerina, I'd need to use my face as much as my feet. I needed to act as well as dance.

That night I stood in front of a mirror and danced the Black Swan. My arms and legs knew what to do. I didn't need to watch them. I focused on my face, especially my eyes. Then I told myself that a girl was trying to steal Skyler away from me.

My face transformed in front of my eyes. My nose flared, my eyes narrowed, and I realized that I looked like Odile. At my next performance, after hundreds of hours of practicing to be the Black Swan, I finally became the Black Swan.

It was my last performance with DTH, and my final performance of the season at the Jacob's Pillow Dance Festival. It was possibly my last performance in the United States for a long time. This performance was an epiphany, and the timing couldn't have been more perfect.

◦• Chapter 36 •◦

TAKING FLIGHT

In December 2012, just six months before the Jacob's Pillow Dance Festival, I had auditioned with Het Nationale Ballet, also known as the Dutch National Ballet. This is one of the top classical companies in the world, so I thought that my chances of being invited to join were slim.

At the end of the audition, Ted Brandsen, the artistic director, came up to me smiling, but he didn't say a word. I was holding my breath. Finally I could no longer hold it. I exhaled loudly and gasped, "Did I make it?"

"Of course. Why would you think that you didn't?" he asked.

How could I answer him? Could I say, "Because I'm black"?

In my mind this invitation was of historical proportions. I felt that the hiring of a very black girl like me by one of the top classical ballet companies was akin to a white man's offering Rosa Parks a seat at the front of the bus in 1955! I happily signed the contract with the Dutch National Ballet in February.

Shortly after making my decision, I took three weeks off from touring with DTH so that I could return to the South African Mzansi Ballet, which had recently been renamed the Joburg Ballet. This time I had no fears. I wasn't returning to a continent that held only terror for me. I was returning to beloved friends, colleagues, and fans.

In March I danced the role of Kitri in the Joburg Ballet's production of *Don Quixote*. Kitri is a joyous and high-energy role. I returned from that performance exhausted but emotionally recharged.

In case I should forget that life is more than ballet, two very special invitations reminded me. I was invited to serve as a volunteer to the United Nations as a spokesperson for children affected by war, and I had the good fortune to be invited to participate in the 2013 Women in the World Conference at the Koch Theater at Lincoln Center. I was videotaped during an interview in which I had a chance to speak about my experiences as a child affected by war.

I opened the event with a dance, and my mother and I were interviewed on the stage about our experiences with international adoption. Then I sat down with her and my sisters as we watched others on that stage talk about their lives. I was deeply moved by the stories of women, many of them born into and still remaining in poverty, who were finding ways to help others.

In late 2012, I was named by the *Huffington Post* in "18 Under 18: *HuffPost Teen*'s List of the Most Amazing Young People of the Year." I didn't feel that I had done anything to deserve this. I couldn't imagine how I had managed to make that list, especially considering that I shared it with such incredible teens, like Malala Yousafzai, the Pakistani teen activist for the education of girls and women who was shot for standing up to the Taliban, and Gabby Douglas, the African American young woman who won a gold medal in the Olympics.

That honor was followed by others when, in 2013, I was named in "Women in the World: 25 Under-25 Young Women to Watch" by the *Daily Beast*, in the *Newsweek* list "125 Women of Impact," and in O's "50 Things that Will Make You Say 'Wow!'" Again, I was aligned with women whose backstories left me breathless with awe and respect for them.

As a result of all these honors and my recent hiring by a classical ballet company, I began to ponder my good fortune. I realized that it didn't start just in the past year. UNICEF estimates that there are three hundred twenty thousand orphans in Sierra Leone, out of a population of approximately six million! Many more children died during the conflict. Only a very small group of children escaped during the height of the war. I was one of the lucky few. That seemed to be the key. Why was I one of the lucky ones?

"Mom, why did you and Daddy adopt? I'm not talking about adopting just me. Why did you adopt in the first place? Wouldn't you have had a lot more money if you only raised Adam and Erik?"

Without a long explanation, my mother simply said, "We were blessed, and with blessing comes responsibility."

Well, I certainly knew that I was blessed. So I supposed that meant I had a responsibility . . . but to do what? That was the mystery that began plaguing me.

"A responsibility to do what?" I asked her.

"To share," Mom answered.

"To share what?"

"You'll have to figure that out on your own," she said.

I tried to think of what I had to share. I didn't have much money to my name, so I couldn't share wealth. I couldn't share my home, because I was already sharing a tiny New York City apartment bedroom with two sisters. I really didn't have anything material to share, other than a closet full of tutus, but I did have a lot of passion, drive, persistence, and hope—especially hope. But how does someone share hope?

When my mother first suggested that I might want to write my memoir and offered to help me do it, I didn't see the point. "I'm only seventeen," I said. "What could I possibly have to share in a memoir?"

Then I realized that I had a responsibility to write the memoir, and I saw what I had to share. In addition to all the rest of my blessings, I had been blessed with a hardy dose of hope. It was hope that enabled me to survive in Africa in the face of abuse, starvation, pain, and terrible danger. It was hope that made me dare to dream, and it was hope that helped that dream take flight. Yes, I would share my hope.

TWO WOMEN

M y life was most profoundly impacted by two women. One is my wonderful mother, whose caring arms I walked into on that June day in 1999, when I arrived at the airport in Ghana.

The second woman is someone I had never met, yet she helped me get through my most terrible days in Sierra Leone and inspired me to be a ballerina. She is the ballerina on the cover of my magazine.

On the night before Mia and I left Africa with our new mother, there had been a blackout in Accra, the capital of Ghana. My mother had packed our bags in pitch-darkness, so she wasn't quite sure where everything was. "My only concern was our money, our passports, and your orphan visas," she later explained.

I was very sick at the airport, so she held me in her arms while she checked our bags and kept track of Mia. Twenty-four hours later, when we arrived at the John F. Kennedy International Airport in New York, we were one bag short. It wasn't until days later that my mother realized that the missing bag contained the picture of the ballerina and the clothes that we had worn in Sierra Leone.

I missed that picture, but eventually real ballet lessons replaced the promise that the picture had held for me. As the years went by, my mother and I searched the Internet for my ballerina. Often Mom would have me pose for her so that she'd know what she was looking for.

One day Mom found her. She was on the cover of *Dance Magazine*. The image we downloaded was only the size of a postage stamp, so we couldn't read the name of the ballerina, and we weren't even sure of the date of publication. When Mom tried to enlarge it, the photo was just a blur, so I kept the tiny image as a memory.

Shortly before I left for Amsterdam to begin my contract with the Dutch National Ballet, a Dutch journalist named Steffie Kouters interviewed me for a magazine. During the interview I told her about the ballerina on the cover of the magazine and what she had meant to me.

Three months later Steffie called my mother and asked for a picture of the cover. Mom apologized for its size and quality, then sent it to her, and Steffie contacted *Dance Magazine*. She was able to get a copy of the cover from the publisher. Now we could see that the issue was dated

May 1979. No wonder the magazine was so worn when I found it.

The ballerina on the cover was Magali Messac, who was a principal dancer with the Pennsylvania Ballet at the time. Coincidentally, that was the same company where I would dance as a Party Girl in *The Nutcracker* twenty-four years later. Eventually, Magali became a principal dancer at ABT, where I trained.

Steffie helped my mother track down Magali. My mother contacted my ballerina and discovered that she was very familiar with my story and had seen *First Position*. She had been moved by the story of the magazine but never dreamt that she was my ballerina. She was deeply touched to learn of the role that she had played in my life, and she said that she wanted to give me her copy of the issue. I cried when my mother emailed me in Amsterdam and told me this.

By the time you read this book, Magali and I will have met. My heart pounds like an African drum at the thought of it. I hope that my mother will be present for our meeting. I would like to have my picture taken with the two women who changed my life for the better.

Acknowledgments

Most of all I want to thank my mother and father, Elaine and Charles DePrince, for their wisdom, sacrifice, love, support, encouragement, and faith in me through the ups and downs of my life. I love you both so much!

I also wish to thank my birth parents for their belief that a girl child is just as good as a boy child, and for believing that girls deserve to go to school.

I thank the real Papa Andrew for sheltering me in his orphanage and bringing me to safety, and the real Uncle Sulaiman, who had the courage to beg for the life of a poor orphan pikin.

Thank you, my dear sister Mia, for the love and friendship we have shared for so many years, in hardship and in joy. Let's not allow time or circumstance to get in the way of that.

I will be eternally grateful to Magali Messac, the ballerina on the cover of my magazine. Her smile and grace gave me hope for my future when I had nothing else.

I want to acknowledge the wonderful people at the

Maine Adoption Placement Service, who worked to bring Mia, Mariel, and me to America, and who continue to help families and rescue children all over the world.

I owe a debt of gratitude to a boy who died before I was born. My brother Michael, for whom I am named, insisted that my parents adopt "a starving orphan from war-torn Africa." Mom and Dad admitted that they might never have thought of it on their own. Michael, I wish I could give you a big hug.

Through my brother Teddy's kindness and generosity of spirit, I was able to overcome my fear of young men. I thank him, and will miss him always. I also thank all of the other members of my family for their love and affection, including brothers, sisters, sisters-in-law, brother-in-law, nieces, and nephew. What a terrific family. I am so happy to have every single one of you in my life.

I want to thank Bo and Stephanie Spassoff, Arthur Mitchell, Mariaelena Ruiz, Charla Genn, Franco De Vita, Kate Lydon, Alaina Albertson-Murphy, Bill Glassman, Susan Jaffe, Natalya Zeiger, Raymond Lukens, and all of the other dedicated ballet teachers who have poured so much energy into my training.

I owe a special debt of gratitude to the artistic directors who had enough faith in me to invite me to dance with their companies. These include Madeline Cantarella Culpo, Virginia Johnson, Dirk Badenhorst, Ted Brandsen, Ernst Meisner, Kevin McKenzie, and my "Dutch papas,"

Rinus Sprong and Thom Stuart, who introduced me to the Netherlands.

I certainly cannot forget Bess Kargman. I'm so glad that you had the persistence to follow me around and the courage to expose issues of race in ballet. I am honored to be a part of *First Position*, which has affected my life in such a positive way.

I am grateful to my friends all over the world who have encouraged me through the years, but especially to the boy who fell in love with me, my beloved Skyler Maxey-Wert.

Mom, again, I can't thank you enough. Not only did you love me, raise me, instill strong values in me, and teach me to cook, but you also taught me to write those five-paragraph essays, and you co-authored this book with me.

And last, but certainly not least, I cannot say thank you enough to the people who believed in this memoir: my literary agent, Adriana Dominguez; my editor, Erin Clarke; and all of the people on my Random House team.

I feel blessed to have had each and every one of you in my life.